A Gallery of Fluid Motion

A Gallery of Fluid Motion

M. Samimy,

The Ohio State University

K. S. Breuer,

Brown University

L. G. Leal,

University of California at Santa Barbara

P. H. Steen

Cornell University

PUBLISHED BY THE PRESS SYNDICATE OF THE UNIVERSITY OF CAMBRIDGE
The Pitt Building, Trumpington Street, Cambridge CB2 1RP, United Kingdom

CAMBRIDGE UNIVERSITY PRESS
The Edinburgh Building, Cambridge CB2 2RU, UK
40 West 20th Street, New York, NY 10011–4211, USA
477 Williamstown Road, Port Melbourne, VIC 3207, Australia
Ruiz de Alarcón 13, 28014 Madrid, Spain
Dock House, The Waterfront, Cape Town 8001, South Africa

http://www.cambridge.org

First published 2003

Printed in the United Kingdom at the University Press, Cambridge

Typeface Times New Roman 10/13 pt *System* QuarkXPress™ [SE]

A catalogue record for this book is available from the British Library

Library of Congress Cataloguing in Publication data

ISBN 0 521 82773 6 hardback
ISBN 0 521 53500 X paperback

Contents

The year in which the winning entry was published in *Physics of Fluids* is given in square brackets.

Introduction ix

1 Jets and mixing layers

M. M. KOOCHESFAHANI AND P. E. DIMOTAKIS
Laser-induced fluorescence measurements of the composition field in a liquid mixing layer [1986] 2

M. GHARIB AND P. DERANGO
Flow studies of a two-dimensional liquid film jet [1988] 3

E. VILLERMAUX, H. REHAB, AND E. J. HOPFINGER
Shear instabilities in the near field of coaxial jets [1998] 4

D. E. PAREKH, W. C. REYNOLDS, AND M. G. MUNGAL
A bifurcating air jet [1987] 5

T. LOISELEUX AND J.-M. CHOMAZ
Breaking of rotational symmetry in a swirling jet experiment [2000] 6

T. F. FRIC AND A. ROSHKO
Views of the transverse jet near field [1988] 7

S. P. GOGINENI, M. M. WHITAKER, L. P. GOSS AND W. M. ROQUEMORE
Dynamics of jet in cross flow [1995] 8

S. WERNZ AND H. F. FASEL
Vortex motion in an unsteady forced wall jet [1996] 9

S. GOGINENI, C. SHIH, AND A. KROTHAPALLI
Comparison of transitional free jet and wall jet [1993] 10

2 Vortices

J. M. LOPEZ AND A. D. PERRY
Periodic axisymmetric vortex breakdown in a cylinder with a rotating end wall [1992] 12

P. KOUMOUTSAKOS AND A. LEONARD
Flow past an impulsively started cylinder [1995] 13

M. P. ESCUDIER
Swirling flow in a cylindrical container [1985] 14

G. J. F. VAN HEIJST, R. C. KLOOSTERZIEL, AND C. W. M. WILLIAMS
Formation of a tripolar vortex in a rotating fluid [1991] 15

H. HIGUCHI AND H. BALLIGAND
Vortex structure behind a disk started from rest [1992] 16

M. VISBAL
Computed unsteady structure of spiral vortex breakdown on a delta wing [1997] 17

G. D. MILLER AND C. H. K. WILLIAMSON
Free flight of a delta wing [1995] 18

H. HIGUCHI
Aircraft trailing vortices and downwash phenomenon [1993] 19

F. M. PAYNE, R. C. NELSON, AND T. T. NG
Vortex breakdown on a delta wing [1986] 20

Q. RAHAMAN, A. ALVAREZ-TOLEDO, B. PARKER, AND C. M. HO
Interaction of two-dimensional wakes [1988] 21

T. LEWEKE, M. PROVANSAL, D. ORMIÈRES, AND R. LEBESCOND
Vortex dynamics in the wake of a sphere [1999] 22

J. C. OWEN, A. A. SZEWCZYK, AND P. W. BEARMAN
Suppression of Kármán vortex shedding [2000] 23

G. J. F. VAN HEIJST AND J. B. FLÓR
Colliding dipolar vortices in a stratified fluid [1990] 24

C. SEREN, M. V. MELANDER, AND N. J. ZABUSKY
Asymmetric vortex merger [1987] 25

P. PETITJEANS
Visualizations of vortex filaments [1998/2002] 26

R. HANCOCK
Visualization of "vaporlines" emanating from water droplets [1992] 27

3 Patterns

G. E. KOPPENWALLNER AND D. ETLING
Vortex flows paint themselves [1992] 29

C.-W. LEONG AND J. M. OTTINO
Mixing in regular and chaotic flows [1989] 30

E. VILLERMAUX AND J. DUPLAT
Mixture's route to uniformity by coalescence [2002] 31

P. D. WEIDMAN, V. O. AFENCHENKO, A. B. EZERSKY, S. V. KIYASHKO, AND M. I. RABINOVICH
The generation of two-dimensional vortices by transverse oscillation of a soap film [2002] 32

E. BODENSCHATZ, S. W. MORRIS, J. R. DE BRUYN, D. S. CANNELL, AND G. AHLERS
Hexagons and spirals in non-Boussinesq convection [1993] 33

C. F. CHEN
Viscosity effects on the directional solidification of a NH_4Cl solution in a Hele-Shaw cell [1992] 34

F. ZOUESHTIAGH AND P. J. THOMAS
Spiral patterns formed by granular media underneath a rotating fluid. Experiment vs. Computation [2000] 35

G. GAUTHIER, P. GONDRET, F. MOISY, AND M. RABAUD
Patterns between two rotating disks [2002] 36

M. FERMIGIER AND P. JENFFER
Sand ripples in a rotating tank [2002] 37

E. TAN AND S. T. THORODDSEN
The thistle crown [1998] 38

B. VUKASINOVIC, A. GLEZER, AND M. K. SMITH
Mode shapes of a sessile drop in forced vibration [2001] 39

N. J. ZABUSKY AND W. TOWNSEND
Naruto: Past and present [1997] 40

R. A. HESS, N. J. BROCK, B. J. WEBER, L. W. CARR, AND M. S. CHANDRASEKHARA
Dynamic stall – a holographic animation of compressible flow interferograms [1992] 41

4 Drops and bubbles

S. CHANDRA AND C. T. AVEDISIAN
The collision of a droplet with a solid surface [1990] 43

M. P. BRENNER, X. D. SHI, J. EGGERS, AND S. R. NAGEL
The bifurcation of liquid drops [1995] 44

M. TJAHJADI AND J. M. OTTINO
Breakup and dispersion of highly stretched droplets in a chaotic flow field [1990] 45

Ph. MARMOTTANT AND E. VILLERMAUX
Ligament mediated drop formation [2001] 46

B. VUKASINOVIC, A. GLEZER, AND M. K. SMITH
Vibration-induced droplet atomization [2000] 47

A. LOZANO, C. J. CALL, AND C. DOPAZO
Atomization of a planar liquid sheet [1994] 48

D. E. NIKITOPOULOS AND A. J. KELLY
Formation of a net-negative charged liquid spray [2000] 49

D. FROST AND B. STURTEVANT
Explosive boiling of a droplet at the superheat limit [1986] 50

M. M. WEISLOGEL AND S. LICHTER
Water balloon rupture in low-g [1998] 51

M. MANGA AND H. A. STONE
Interactions between bubbles at low Reynolds numbers [1993] 52

J. BUCHHOLZ, L. SIGURDSON, AND B. PECK
Bursting soap bubble [1995] 53

5 Complex fluids

I. ARANSON, D. BLAIR, AND P. VOROBIEFF
Interface motion in a vibrated granular layer [1999] 55

G. METCALFE, T. SHINBROT, J. J. MCCARTHY, AND J. M. OTTINO
Mixing patterns for surface dominated granular flows [1995] 56

J. S. OLAFSEN AND J. S. URBACH
Crystallization in a 2D granular fluid [1999] 57

R. MIKKELSEN, M. VERSLUIS, E. KOENE, G.-W. BRUGGERT, D. VAN DER MEER, K. VAN DER WEELE, AND D. LOHSE
Granular eruptions: Void collapse and jet formation [2002] 58

M. TIRUMKUDULU, A. TRIPATHI, AND A. ACRIVOS
Particle segregation in monodisperse sheared suspension [1999] 59

J. H. WALTHER, S.-S. LEE, AND P. KOUMOUTSAKOS
Simulation of particle laden flows using particle methods [2000] 60

I. EAMES AND S. B. DALZIEL
Resuspension by an impacting sphere [1999] 61

S. L. ANNA, H. SPIEGELBERG, AND G. H. MCKINLEY
Elastic instability in elongating fluid filaments [1997] 62

6 Flows with interfaces

A. E. HASHA AND J. W. M. BUSH
Fluid fishbones [2002] 64

M. F. G. JOHNSON, M. J. MIKSIS, R. A. SCHLUTER, AND S. G. BANKOFF
Fluid chains produced by obliquely intersecting viscous jets connected by a thin free liquid film [1996] 65

I. L. KLIAKHANDLER, S. H. DAVIS, AND S. G. BANKOFF
Viscous beads on thin vertical fiber [2001] 66

B. J. FISCHER, A. A. DARHUBER, AND S. M. TROIAN
Front propagation and branching dynamics in surfactant driven spreading [2001] 67

T. MAXWORTHY
A note on the nonlinear growth of a gravitationally unstable interface in a Hele-Shaw cell [1985] 68

R. BUCKINGHAM AND J. W. M. BUSH
Fluid polygons [2001] 69

S. T. THORODDSEN AND L. MAHADEVAN
Shark-teeth pattern in coating flow inside a horizontally rotating cylinder [1996] 70

A. A. VEDERNIKOV, B. SCHEID, E. ISTASSE, AND J. C. LEGROS
Viscous fingering in miscible liquids under microgravity conditions [2001] 71

7 Free surface interaction

M. ERRICO
Laminar jets can splash! [1986] 73

B. PECK AND L. SIGURDSON
Impacting water drops [1992] 74

Y. ZHU, H. N. OĞUZ, AND A. PROSPERETTI
Air cavity due to jet disturbance [1998] 75

B. W. ZEFF, J. FINEBERG, AND D. P. LATHROP
Formation of a self-focusing singularity on a fluid surface [1999] 76

D. CHIRICHELLA, R. GÓMEZ LEDESMA, K. KIGER, AND J. H. DUNCAN
Air entrainment by a plunging jet translating over a free surface [2002] 77

L. SIGURDSON
Atom bomb/water drop [1991] 78

L. L. SHAW, S. A. MUELDER, D. W. BAUM, AND K. A. WINER
Hypervelocity explosive-driven metal jet in air [1994] 79

D. LIEPMANN AND M. GHARIB
The effect of the free surface on a round jet [1990] 80

8 Combustion

A. YOSHIDA
Wrinkled laminar flames [1989] 82

W. M. ROQUEMORE, L. D. CHEN, J. P. SEABA, P. S. TSCHEN, L. P. GOSS, AND D. D. TRUMP
Jet diffusion flame transition to turbulence [1987] 83

K. Y. HSU, V. R. KATTA, L. P. GOSS, D. D. TRUMP, L. D. CHEN, AND W. M. ROQUEMORE
Flame-vortex interactions in a driven diffusion flame [1993] 84

E. PFIZENMAIER, J. SIMON, AND P. A. MONKEWITZ
Bouquet with bifurcating jet diffusion flame [1993] 85

N. ASHGRIZ
Flame impingements [1989] 86

T. ECHEKKI AND M. G. MUNGAL
Particle tracking in a laminar premixed flame [1990] 87

9 Instability

W. M. ROQUEMORE, R. L. BRITTON, R. S. TANKIN, C. A. BOEDICKER, M. M. WHITAKER, AND D. D. TRUMP
Interaction of 2D wake and jet plume [1988] 89

T. LEWEKE AND C. H. K. WILLIAMSON
The long and short of vortex pair instability [1996] 90

C. E. NIEDERHAUS AND J. W. JACOBS
Instability of an impulsively accelerated liquid/liquid interface [1998] 91

M. R. RUITH AND E. MEIBURG
Breakdown modes of swirling jets with coflow [2002] 92

P. M. RIGHTLEY, P. VOROBIEFF, AND R. F. BENJAMIN
Mushrooms and snakes: A visualization of Richtmyer-Meshkov instability [1997] 93

J. M. SKOTHEIM AND J. W. M. BUSH
Evaporatively driven convection in a draining soap film [2002] 94

Y. COUDER
Anomalous Saffman-Taylor fingering [1987] 95

C. D. ANDERECK AND H. L. SWINNEY
Flows in the circular Couette system [1985] 96

10 Transition and turbulence

A. PRASAD AND C. H. K. WILLIAMSON
A new mechanism for oblique wave resonance [1994] 98

W. S. SARIC
Visualization of different transition mechanisms [1986] 99

T. PEACOCK AND T. MULLIN
The transition to turbulence in a microscopic fluid flow [2000] 100

A. DRAKE, R. V. WESTPHAL, R. A. KENNELLY, JR., AND D. M. DRIVER
Interferometric skin friction footprint for a tripped boundary layer [1998] 101

J. H. DUNCAN, V. PHILOMIN, H. QIAO, AND J. KIMMEL
The formation of a spilling breaker [1994] 102

M. A. RUTGERS, X.-L. WU, AND W. I. GOLDBURG
The onset of two-dimensional grid generated turbulence in flowing soap films [1996] 103

G. ZOCCHI, E. MOSES, AND A. LIBCHABER
Coherent structures in thermal turbulence [1991] 104

D. R. SABATINO AND T. J. PRAISNER
The colors of turbulence [1998] 105

S. GOGINENI, R. RIVIR, D. PESTIAN, AND L. GOSS
High free-stream turbulence influence on turbine film cooling flows [1996] 106

Y.-B. DU AND P. TONG
Turbulent thermal convection over a rough surface [2000] 107

11 Compressible flows

R. DARKE AND J. B. FREUND
Mach wave radiation from a jet at Mach 1.92 [2001] 109

T. A. MANNING AND S. K. LELE
Sound due to the interaction of shear-layer instability-waves and an oblique shock [1999] 110

B. YIP, K. LYONS, M. LONG, M. G. MUNGAL, R. BARLOW, AND R. DIBBLE
Visualization of a supersonic underexpanded jet by planar Rayleigh scattering [1989] 111

J. W. JACOBS
PLIF flow visualization of a shock-accelerated light gas cylinder [1989] 112

J. F. HAAS AND B. STURTEVANT
Shock-induced deformation and mixing of a helium sphere immersed in air [1986] 113

M. WATANABE AND K. TAKAYAMA
Converging cylindrical shock wave [1990] 114

H. KLEINE AND K. TAKAYAMA
Combined schlieren and interferometry visualization of blast waves [2001] 115

Keyword Index 116

Introduction

The desire to capture images of fluids in motion for both scientific and artistic reasons dates back over 500 years, at least to the time of Leonardo de Vinci who is known to have recorded visual images of the complicated patterns traced by floating seeds on the surface of naturally flowing water. Indeed, flow visualization has played a major role in the development of the science of fluid mechanics, and has certainly been a key component in major technological advances such as the evolution of flight; in understanding natural phenomena such as the atmospheric motions that are responsible for weather-related phenomena such as hurricanes or tornadic thunderstorms; and in understanding biological systems such as the heart pump. A collection of some of the most striking photographs of fluid motions from the research literature was collected and published in 1982 in a book entitled *An Album of Fluid Motion*, by Milton Van Dyke.

Motivated by Van Dyke's book, the Division of Fluid Dynamics (DFD) of the American Physical Society (APS) has sponsored a "photo" contest at its annual scientific meeting each year since 1983. Researchers are invited to display visual images of fluids in motion. The entries are judged by a distinguished panel of fluid dynamics researchers to choose the most outstanding contributions based upon two criteria:

(1) The artistic beauty and novelty of the visualizations;
(2) The contribution to a better understanding of fluid flow phenomena.

Since 1986, the winning entries have been published in a series of articles called the "Gallery of Fluid Motion" that is published each year in the scientific journal, *Physics of Fluids*. Offprints of this article are distributed to participants at the next Annual Meeting of the APS-DFD.

This book collects images of fluid flow phenomena selected from the "Gallery of Fluid Motion." The selections have been edited according to the above criteria and assembled to be accessible to a broad community.

Although originally conceived as an annual fluid mechanics photography contest, the annual picture gallery rapidly evolved to become a display of all types of visual imaging of fluid motions, including not only photographic visualizations from laboratory experiments, but also digitally generated visualizations from computational simulations of flow phenomena, and animated visualizations based first on video tapes (1986) and more recently on computer-based imaging programs. The winning animated visualizations were replaced by still photo displays for the printed "Gallery" article, and some of those appear in this book. Since 2000, however, the video images have also been published in the online version of the "Gallery" article, and are freely accessible on the Web (http://ojps.aip.org/phf/gallery/).

The beauty and natural complexity of the phenomena displayed each year in the annual picture gallery have educated, entertained, and generally been a highlight for the scientific community who attend the Annual DFD-APS Meeting. New phenomena are often displayed for the first time, and there is much to be learned about important and complex phenomena from all of the entries to the picture gallery. Equally important, however, is the feast to the senses that is contained in these often striking presentations. Until now, however, an easily accessible collection of these flow-visualization images has not been available outside our scientific community.

The editors of this book are all active participants in the DFD-APS, and are acting on behalf of the Executive Committee as well as the general membership of the DFD, to produce a collection of the "best" images from the original "Gallery of Fluid Motion" articles. Approximately 60% of the original winning entries have been selected by us to appear in this book, again based primarily on the above-mentioned two criteria. We then categorized the selected images into the broad sub-sections of fluid mechanics that are indicated in the Table of Contents. We recognize that these designations are somewhat arbitrary in that a given entry could often have been incorporated into more than one of these main headings. In such cases, our goal was to keep complementary images of similar phenomena in a single section. There are, of course, many wonderful images in the annual "Gallery of Fluid Motion" articles that were not selected, or for other reasons, mentioned below, could not be used here. One major contributory factor in the selection process was that early entries were often superseded by later entries showing the same or similar phenomena, but with the advantage of color imaging or better visualization methods.

A few words are probably needed to explain some of the editorial issues with this book, especially in the present electronic world. Much to our surprise, we discovered that Physics of Fluids only had electronic archives from 2000 forward, and had returned all of the original photos, etc., to the authors over the years. Hence, after making selections for what would be in the book, we had to find and contact the authors to get all the originals, which turned out to be quite a task. In some cases we borrowed framed original photographs that were decorating the office of one of the authors to reproduce them. Fortunately, there were only a few cases where either we could not locate the authors, or the original photos were not available. In a small number of cases, the images in this book are slightly different or updated versions of the original photos that appeared in Physics of Fluids. Finally, since this book represents reproduced images of published materials, we felt that it was most appropriate to retain the original author affiliations, even though one or more of the authors may have subsequently relocated.

We hope that the collection of images that we have put together will be of interest both to the research specialist and to the general public. In all cases, we have left the original description of the image as it appeared in the original "Gallery of Fluid Motion" article, except for minor updating. Although these descriptions were written originally for the research community, we hope that all readers will enjoy the beauty of the images we have selected and that many of these will be self-explanatory in relation to the phenomena that they illustrate. At the same time, we hope that the written descriptions will provide the needed information for any reader who wishes to ponder more deeply the physical significance of the flow visualizations, or who might wish to continue research into the phenomena that are displayed.

1 Jets and mixing layers

Laser-induced fluorescence measurements of the composition field in a liquid mixing layer

M. M. Koochesfahani and P. E. Dimotakis
California Institute of Technology

Laser-induced fluorescence (LIF) diagnostics and high-speed, real-time digital image acquisition techniques are combined to map the composition field in a water mixing layer.[1] A fluorescent dye, which is premixed with the low-speed freestream fluid and dilutes by mixing with the high-speed fluid, is used to monitor the relative concentration of high-speed to low-speed fluid in the layer.

The three digital LIF pictures shown here were obtained by imaging the laser-induced fluorescence originating from a collimated argon ion laser beam, extending across the transverse dimension of the shear layer, onto a 512-element linear photodiode array. Each picture represents 384 contiguous scans, each at 400 points across the layer, for a total of 153 600 point measurements of concentration. The vertical axis maps onto 40 mm of the transverse coordinate of the shear layer, and the horizontal axis is time increasing from right to left for a total flow real time of 307 msec. The pseudocolor assignment is linear in the mixture fraction (ξ) and is arranged as follows: red–unmixed fluid from the low-speed stream ($\xi = 0$); blue–unmixed fluid from the high-speed stream ($\xi = 1$); and the rest of the spectrum corresponds to intermediate compositions.

Figures 1 and 2, a single vortex and pairing vortices, respectively, show the composition field before the mixing transition. The Reynolds number based on the local visual thickness of the layer and the velocity difference across the layer is Re = 1750 with $U_2/U_1 = 0.46$ and $U_1 = 13$ cm/sec. Note the large excess of high-speed stream fluid in the cores of the structures.

Figure 3 shows the composition field after the mixing transition. The horizontal scale is reduced by a factor of about 5 compared to Figures 1 and 2. The Reynolds number is Re = 23 000 with $U_2/U_1 = 0.38$ and $U_1 = 70$ cm/sec. Note that the composition of the mixed fluid is asymmetric and biased toward the high-speed fluid and that it is almost uniform across the entire transverse extent of the shear layer.

Keywords
vortex pairing; free shear layer; mixing transition; laser-induced fluorescence.

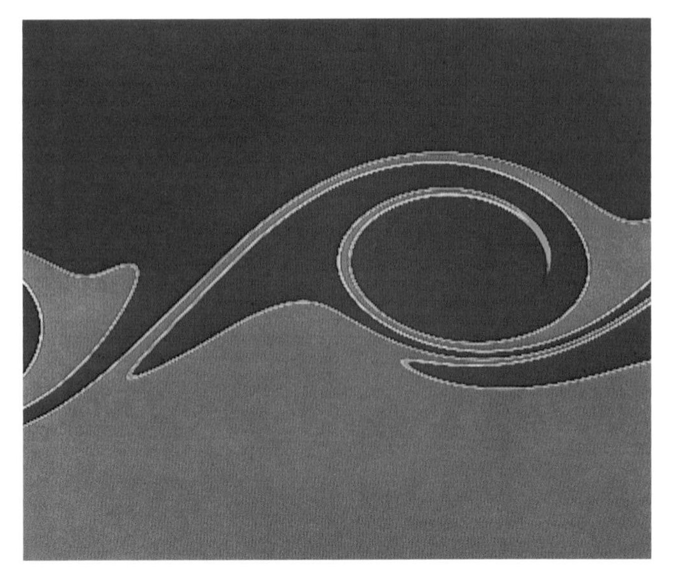

Figure 1

Figure 2

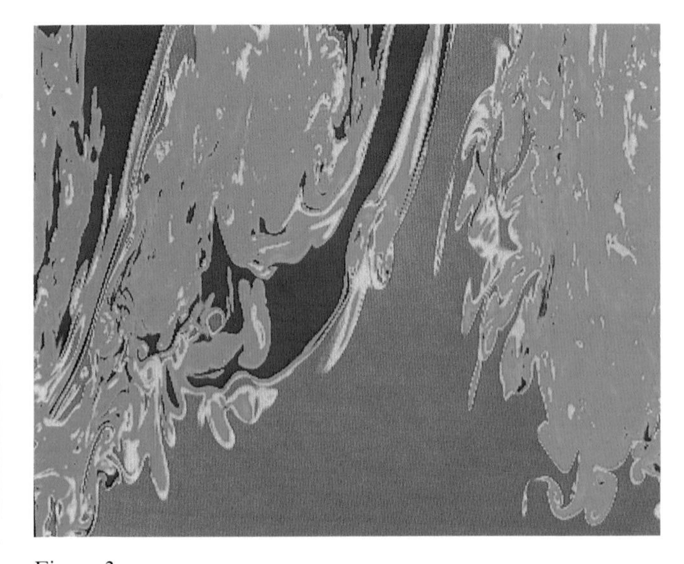

Figure 3

[1] M. M. Koochesfahani and P. E. Dimotakis, *AIAA J.* **23**, 1700 (1985).

Figure 1

Figure 2

Flow studies of a two-dimensional liquid film jet

M. Gharib and P. Derango

University of California, San Diego

The above images represent a two-dimensional jet produced in a soap film tunnel. The tunnel consists of a frame in which one end is positioned in a diluted soap mixture and the other end is subjected to a film-pulling mechanism. In our device the pulling mechanism is provided through the contact action of a two-dimensional water jet. The two-dimensional jet of higher surface tension is directed at a small angle to the soap film surface at the downstream end of the tunnel. The pulling effect of the high momentum jet results in a uniform two-dimensional motion of the suspended soap film in the frame. Once the two-dimensional flow of the thin film ($\sim$1 μ thick) starts in the frame, various objects can be placed in the test segment of the frame to study their associated two-dimensional flow fields. By imposing certain geometries on the boundaries of the frame, various shear flows such as jets or two-dimensional mixing layers can be produced.

The interference colors in the above images represent isovelocity (isothickness) regions. The two-dimensional jet in Fig. 1 has a centerline velocity of 140 cm/sec with a corresponding Reynolds number of 2000. Figure 2 shows the same jet, but the jet fluid has a lower surface tension than the ambient fluid, which results in a large growth rate for the jet.

Keywords

soap film.

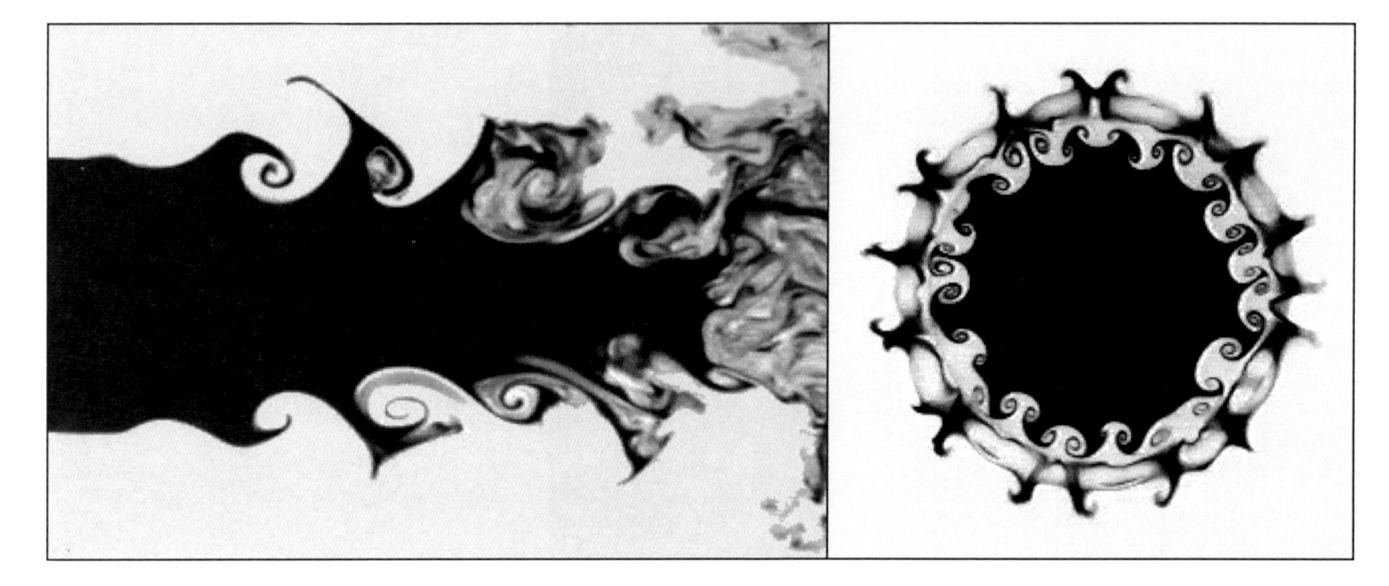

Figure 1

Shear instabilities in the near field of coaxial jets

E. Villermaux, H. Rehab, and E. J. Hopfinger

LEGI-CNRS, Institut de Mécanique de Grenoble, BP 53X, 38041 Grenoble Cedex, France

Snapshots are shown of the destabilization of a slow round jet (velocity u_1) by a fast coflowing annular jet (velocity u_2) in water with $u_2/u_1=3$ as manifested by the mixing of a high Schmidt number dye. The Reynolds number based on the outer diameter and velocity $Re=u_2D_2/\nu$ is 2×10^4. The wavelength $\lambda_\|$ of the longitudinal shear instability at the interface between the two streams (right picture, outer jet seeded) is about six times the vorticity thickness δ of the fast stream velocity profile at the lip of the annular tube, and the wavelength in the transverse direction $\lambda_\perp$ (left picture, transverse cut perpendicular to the direction of the flow at one inner diameter D_1 downstream of the exit plane) is about 3δ.

The development of the longitudinal and transverse instabilities is nearly concomitant and they both grow at the same rate. The longitudinal vortices from the transverse instability (mushroom-like structures on the right picture) connect rapidly the injection scale with the dissipative scale, resulting in an efficient mixing in the near field. The overall entrainment process is completed at a distance $[6/(u_2/u_1)]D_1$ downstream of the injection plane.[1]

When the velocity ratio is further increased beyond about 8, a transition to an unsteady, wake-type recirculation regime is observed.[1,2,3] The recirculation bubble oscillates periodically with a low frequency f, distinct from the jet mode, characterized by a Strouhal number based on the inner diameter and the outer velocity fD_1/u_2 of the order of 0.035.

[1] E. Villermaux, H. Rehab, and E. J. Hopfinger, "Breakup regimes and self-sustained pulsations in coaxial jets," *Meccanica* **29**, 393 (1994).

[2] Rehab, E. Villermaux, and E. J. Hopfinger, "Flow regimes of large velocity ratio coaxial jets," *J. Fluid Mech.* **345**, 357 (1997).

[3] E. Villermaux and H. Rehab, "Mixing in coaxial jets," *J. Fluid Mech.* **425**, 161–185, (2000).

Keywords

round jet; mushroom vortex; shear layer; dye visualization.

A Bifurcating air jet

D. E. Parekh, W. C. Reynolds, and M. G. Mungal
Stanford University

The two round jets pictured are an example of flow control through acoustic excitation.[1] A loudspeaker in the upstream plenum generates a sinusoidal fluctuation in the streamwise direction. Two speakers adjacent to the jet exit produce a transverse perturbation.

Ordinary smoke and an ordinary 35 mm camera are combined with an extraordinary light source to capture the remarkable structures in this flow. Cigar smoke marks the shear layer and vortex rings of the jet. A pulsed light sheet from a copper-vapor laser illuminates the jet cross section. Though the jet exits the nozzle at 7 m/sec, all motion is frozen by the short pulse duration (25 nsec).

The shear layer of a turbulent jet typically rolls up into distinct vortex rings which subsequently interact to form larger structures. In an unexcited jet, vortices form and pair in a somewhat random manner. By streamwise excitation one can control the vortex formation process (Fig. 1). Properly combining streamwise and transverse excitations causes the jet to split into distinct jets (Fig. 2). This phenomenon, known as the bifurcating jet, occurs when the frequency ratio of streamwise-to-transverse excitations is 2. The Reynolds and Strouhal numbers based on diameter are 10 000 and 0.6 in both cases.

The work was sponsored by the Air Force Office of Scientific Research.

Keywords
acoustic forcing; free shear layer; vortex dynamics; smoke visualization.

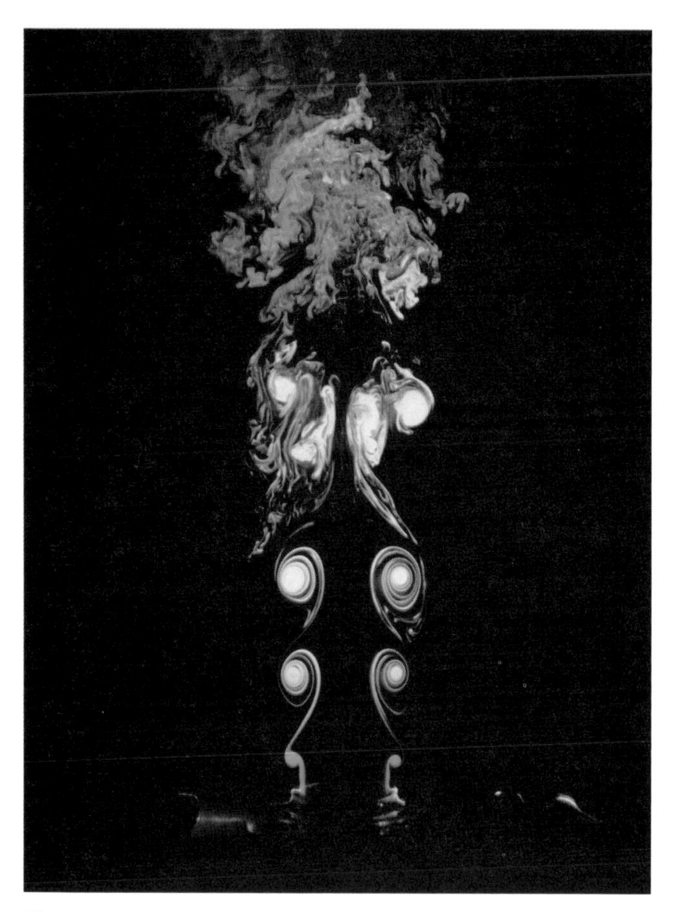

Figure 1

Figure 2

[1] W. C. Reynolds, D. E. Parekh, P. J. D. Juvet, M. J. D. Lee, "Bifurcating and Blooming Jets," *Ann. Rev. Fluid Mech.* **35**, 295–315 (2003).

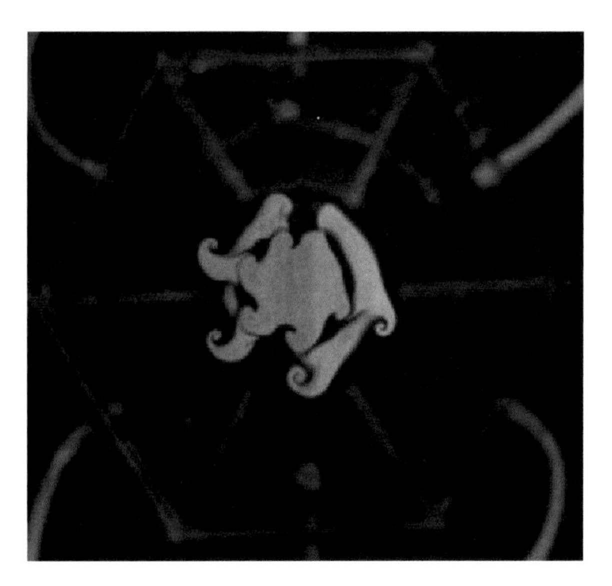
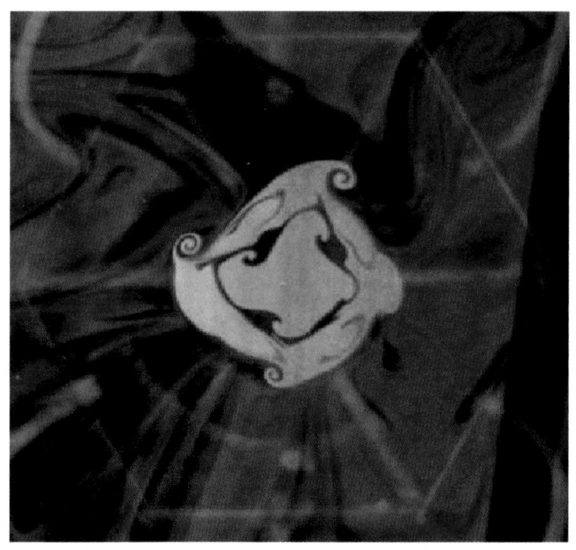

Figure 1 Cross-sectional images at $S = 0.38$ (left picture) and $S = 0.49$ (right picture) of a counter-clockwise swirling jet. In the left picture, *six* intense azimuthal rolled-up structures (four in the right picture) are developing. In the absence of a swirl (not shown), *seven* weak azimuthal deformations are identifiable, which develop further downstream into mature mushroom-like structures composed of counter-rotating streamwise vortex pairs: rotation enhances the concentration of braid vorticity into secondary streamwise structures. Note that azimuthal roll-up is of one sign, thereby indicating that streamwise vortices are all co-rotating with respect to each other, and anticyclonic with respect to the base flow rotation. By contrast, in the nonswirling case, streamwise vortices are counter-rotating pairs. Re = 1490, $X = 3$, where the Reynolds number is based on the nozzle exit diameter D, and X denotes the nondimensional distance x/D to the nozzle exit.

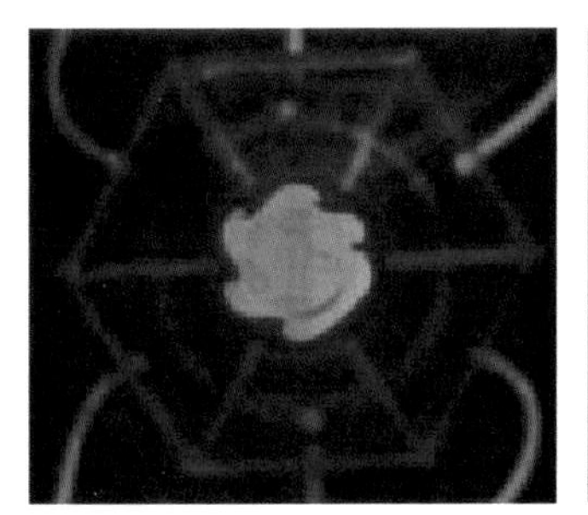
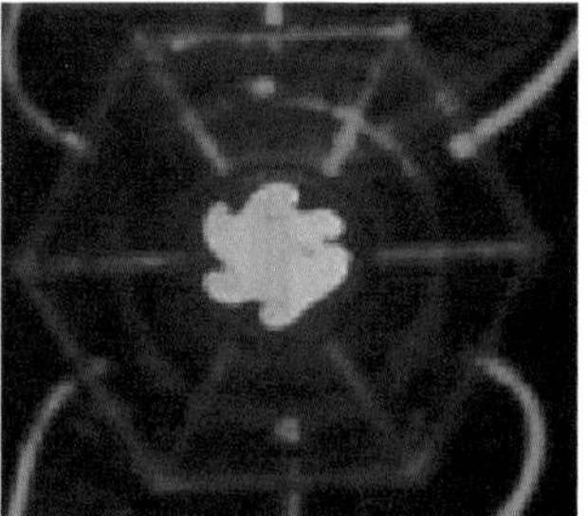
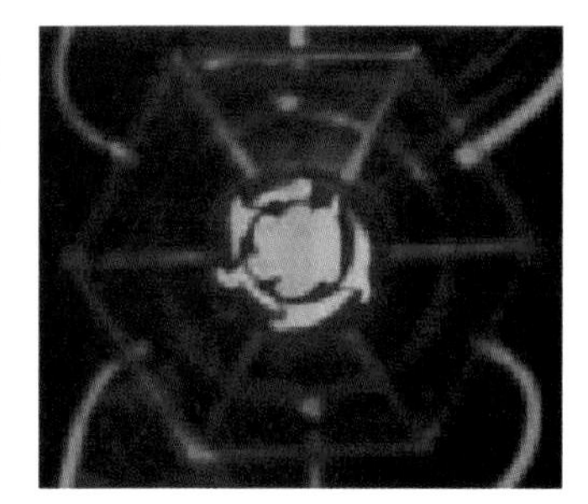
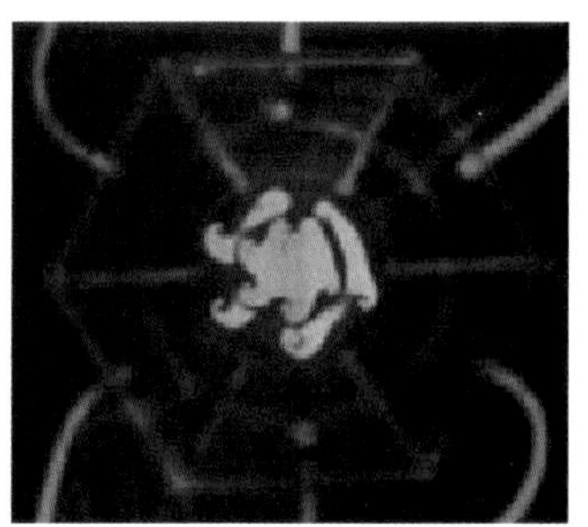

Figure 2 This time sequence of cross-sectional views in the braid (first two pictures) and in the ring (last two pictures) at $S= 0.38$ reveals the advection by the mean rotation of the secondary structures. This entrainment generates an azimuthal wave propagating cyclonically when compared to the imposed rotation, at an angular phase velocity proportional to the swirl. Re = 1490, $X = 3$.

Breaking of Rotational Symmetry in a Swirling Jet Experiment

Thomas Loiseleux and Jean-Marc Chomaz

CNRS–École polytechnique

The symmetry-breaking instabilities in swirling jets are analyzed experimentally for low swirl parameter S, where S compares the magnitudes of the azimuthal and axial velocity components. *Without swirl*, the axial shear generates vortex rings, whereas counter-rotating streamwise vortex pairs form in the braids connecting the rings due to a secondary instability. *For low swirl*, $0 \leq S \leq 0.6$, we have determined that the underlying dominant instability mechanisms are direct extensions of those prevailing in classical nonrotating jets. However, significant differences appear in the development of the secondary instability: Swirl does not affect, qualitatively, the formation of vortex rings, it strengthens the development of azimuthal deformations and causes a decrease in the azimuthal wavenumber m. Finally, swirl amplifies anticyclonic streamwise vortices and attenuates cyclonic streamwise vortices.

We gratefully acknowledge A. Garcia and P. Lavialle for their kind assistance.

Keywords

vortex ring; vortex braids.

Figure 1

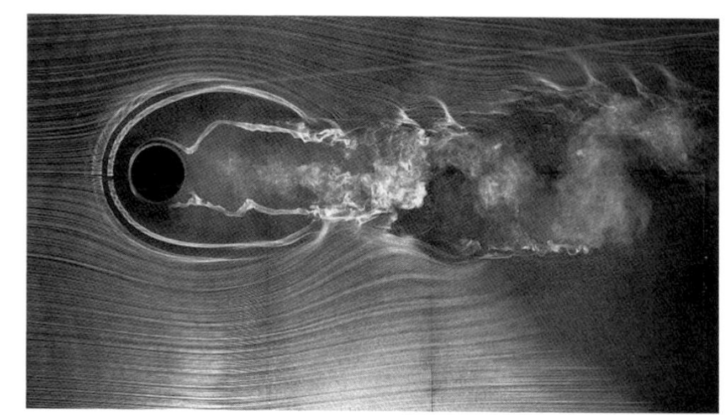

Figure 2

Figure 3

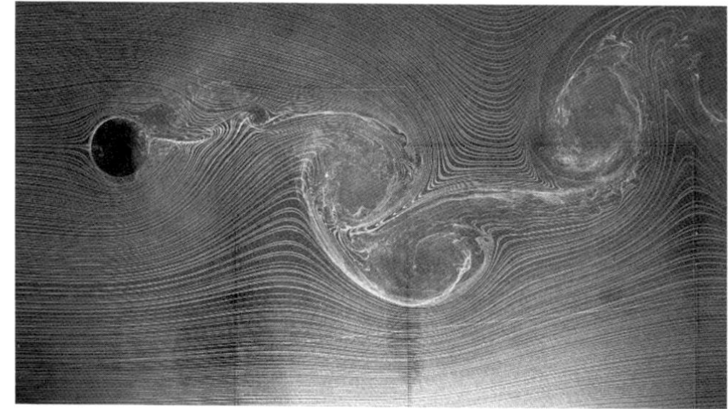

Figure 4

Views of the transverse jet near field

T. F. Fric and A. Roshko

California Institute of Technology

The wide variety of three-dimensional structure in the near field of the transverse jet flow is exemplified by the photographs above. The experimental setup is a 1.5 in. air jet mounted flush with a false side wall of a wind tunnel. The smoke-wire flow visualization technique with flash lighting is used to visualize the dominant near field vortical structures. These structures include the following: distorted ring vortices comprising the shear layer of the bending jet, a system of horseshoe vortices near the side wall, and a system of wake vortices.

Figure 1 shows the rollup of vorticity along the distorted circumference of the bending jet. The jet is issuing from the top, and the cross-flow is from left to right. The reference jet to cross-flow velocity ratio U_{jet}/U_{cf} is 2 and the Reynolds number based on the jet diameter and velocity R_{jet} is 7600. The approaching smoke lines are in the centerplane of the jet, but the smoke is seen coming out of that plane, as a result of axial flow along the cores of the distorted vortex rings.

In Fig. 2 a horseshoe vortex system is visualized with smoke in the cross-flow laminar boundary layer, whose displacement thickness is approximately 0.1 in. The jet, at the left side of the photograph, issues directly at the viewer. Here $U_{jet}/U_{cf}=2$ and $R_{jet}=15\,200$. Three horseshoe vortices are seen to wrap around the jet. Such a horseshoe vortex system is somewhat similar to what is observed in the near wall flow around a wall-mounted circular cylinder. In Fig. 1, cross sections of two horseshoe vortices are visible just upstream of the jet.

Figures 3 and 4 offer two views of the type of vortices observed in the wake of the transverse jet. Figure 3 offers a side view of the wake vortices. The smoke wire is positioned in the jet's wake and aligned with the transverse axis. The jet is five jet diameters upstream of the field of view and issues from the top. Here $U_{jet}/U_{cf}=10$ and $R_{jet}=76\,200$. As another example, Fig. 4 shows a cross-sectional view of the wake vortices, with the jet issuing at the viewer. The velocity ratio $U_{jet}/U_{cf}=4$ and $R_{jet}=45\,700$. The smoke wire is upstream and is placed 0.5 jet diameters from the side wall. Note also the "footprints" of wake vortices observed downstream of the jet in Fig. 2.

This work is sponsored by the Office of Naval Research.

Keywords

horseshoe vortex; vortex ring; wake vortex; smoke visualization.

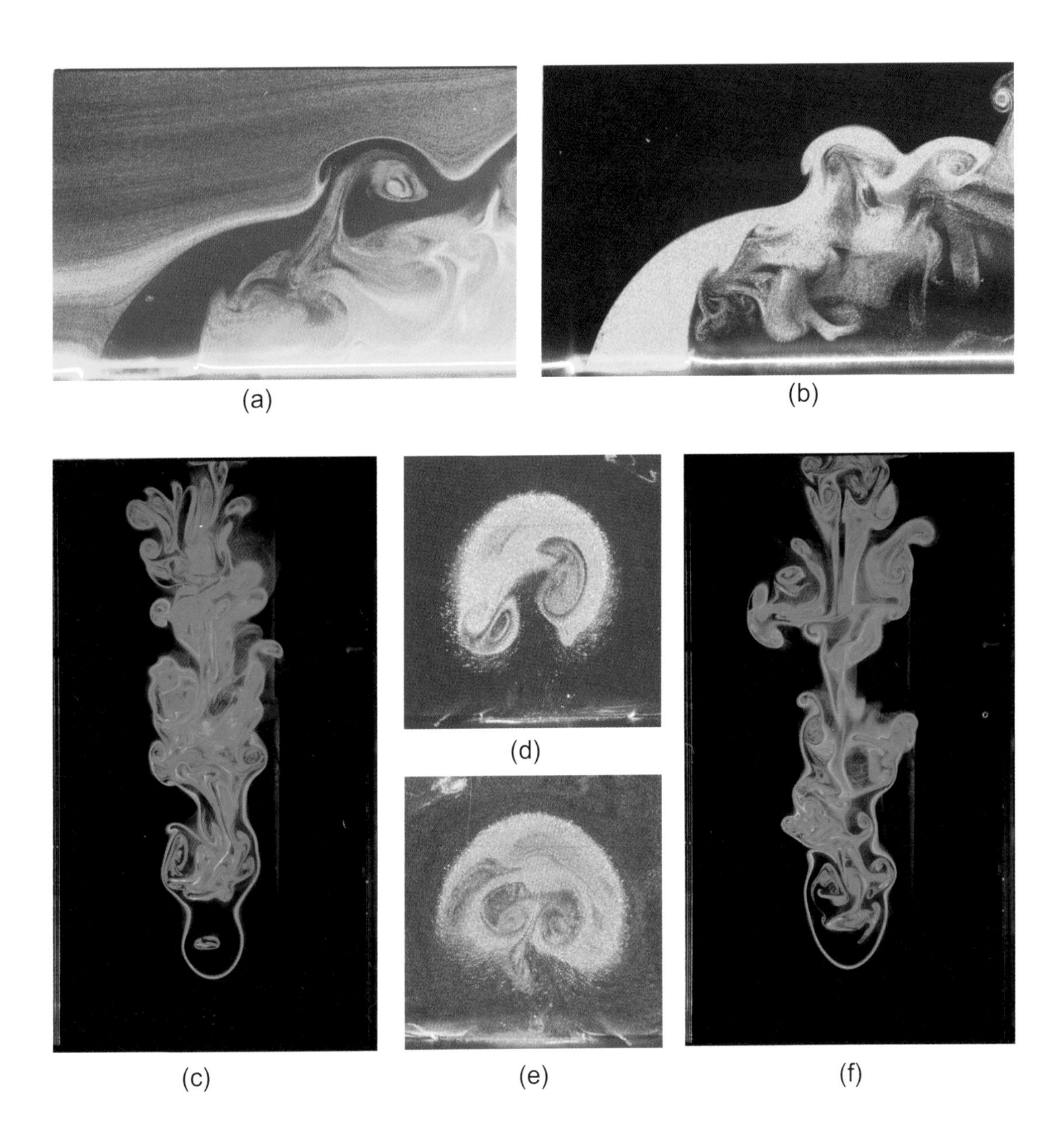

Dynamics of jet in cross flow

S. P. Gogineni, M. M. Whitaker and L. P. Goss, and M. Roquemore

Systems Research Laboratories, Dayton, OH Air Force Research Laboratory, WPAFB, OH

The Mie scattering technique is used to visualize the jet in a cross flow. The jet is issued from a square hole having an equivalent diameter of 1.72 cm. The jet-to-cross flow dynamic head is 1.0. The Reynolds numbers for the jet and the cross flow are 700 and 8775, respectively. The images are acquired at two different instants of time by seeding the flow with tracer particles such as 1-μm aluminum-oxide particles, sub-micron size smoke particles and illuminating them with a two-color laser system. The reactive Mie scattering images [Figs. (c) and (f)] provide information concerning the mixing between the two fluids. The top images [Figs. (a) and (b)] show the penetration of the jet flow into the cross-flow region and the entrainment of the cross-flow fluid into the wake region of the jet. Figures (d) and (e) the counter-rotating vortex pair and its dynamics.

Keywords

vortex pairing; Mie scattering.

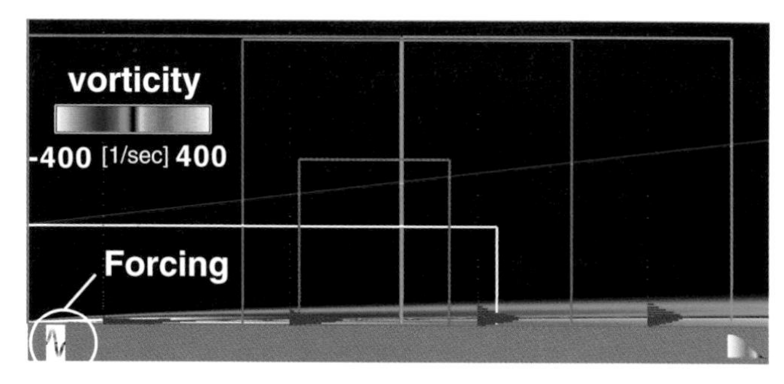

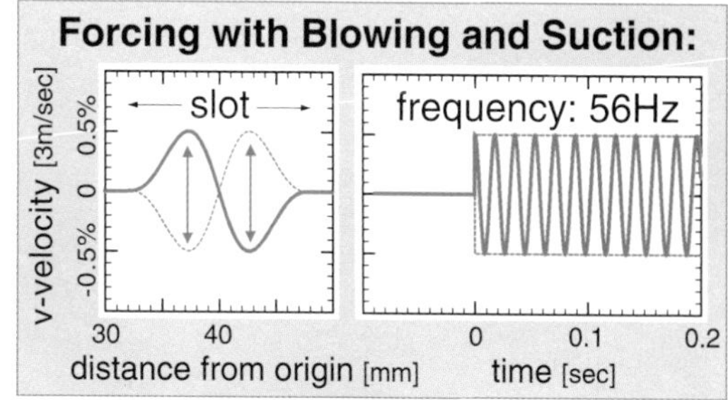

Figure 1 (above) Computational domain with base flow (Glauert's similarity solution) and method of forcing.

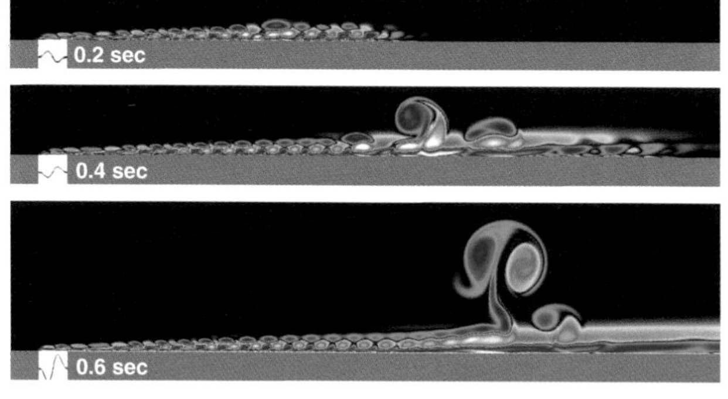

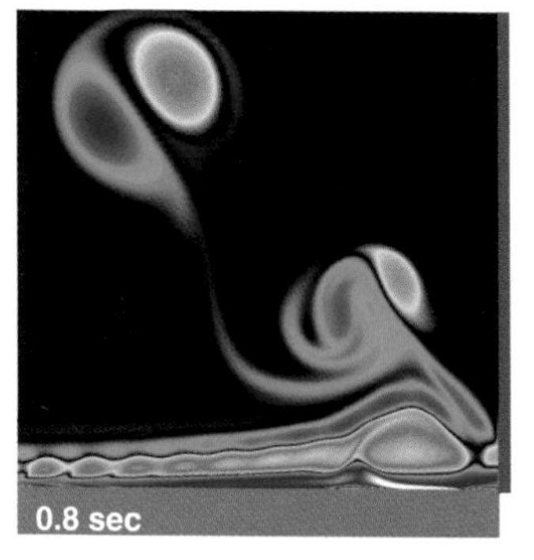

Figure 2 Time sequence for the disturbance flow during startup of forcing with blowing and suction.

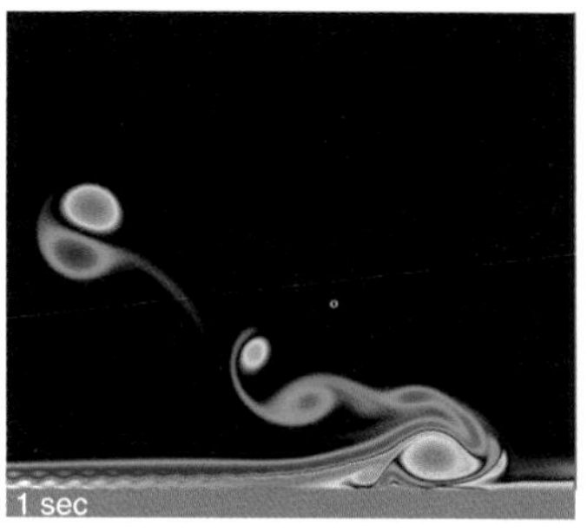

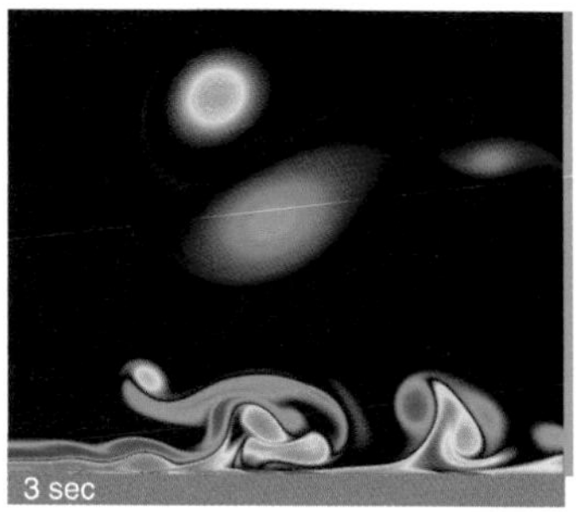

Figure 3 Snapshots of the flow (base flow and disturbance flow) at three instants in time (1, 3, and 10 s).

Vortex motion in an unsteady forced wall jet

Stefan Wernz and Hermann F. Fasel
University of Arizona

Transitional wall jets are being investigated[1] using direct numerical simulations (DNS). In the present simulation, the two-dimensional vorticity equation in disturbance flow formulation is solved numerically over the computational domain shown in Fig. 1 ($Re_{\delta 0.5}$=277–701). The base flow for the computation is given by Glauert's similarity solution, as shown in Fig. 1, using color contours of spanwise vorticity. By forcing the wall jet with periodic blowing and suction through a slot in the wall (Fig. 1) traveling disturbance waves are introduced into the flow. During startup of periodic forcing with large amplitudes, a series of mergings of subsequent vorticity concentrations occurs within the disturbance waves (Fig. 2). These mergings are manifestations of a subharmonic resonance between the fundamental disturbance wave (f= 56 Hz) and disturbance waves with lower frequencies that are generated momentarily by the sudden startup of forcing. This secondary instability process leads to an accumulation of vorticity within one pair of vorticities that is eventually ejected from the wall jet layer into the ambient fluid. By mutual induction this mushroom shaped vortex pair travels upstream and then interacts repeatedly with the wall jet. As a consequence, a very complex flow pattern develops (Fig. 3) that is sensitive to initial conditions and, in the present two-dimensional calculation, persists for about 30 s before the flow reaches a time-periodic state. In three-dimensional computations, breakdown to turbulence takes place shortly after the first vortex pair is ejected.

This research is funded by AFOSR under Contract No. F49620–94–1–0208.

[1] S. Wernz and H. F. Fasel, "Numerical investigation of unsteady phenomena in wall jets," 34th Aerospace Sciences Meeting and Exhibit, *AIAA* Paper No. 96–0079, 1996.

Keywords
simulation; resonance forcing; vortex ejection.

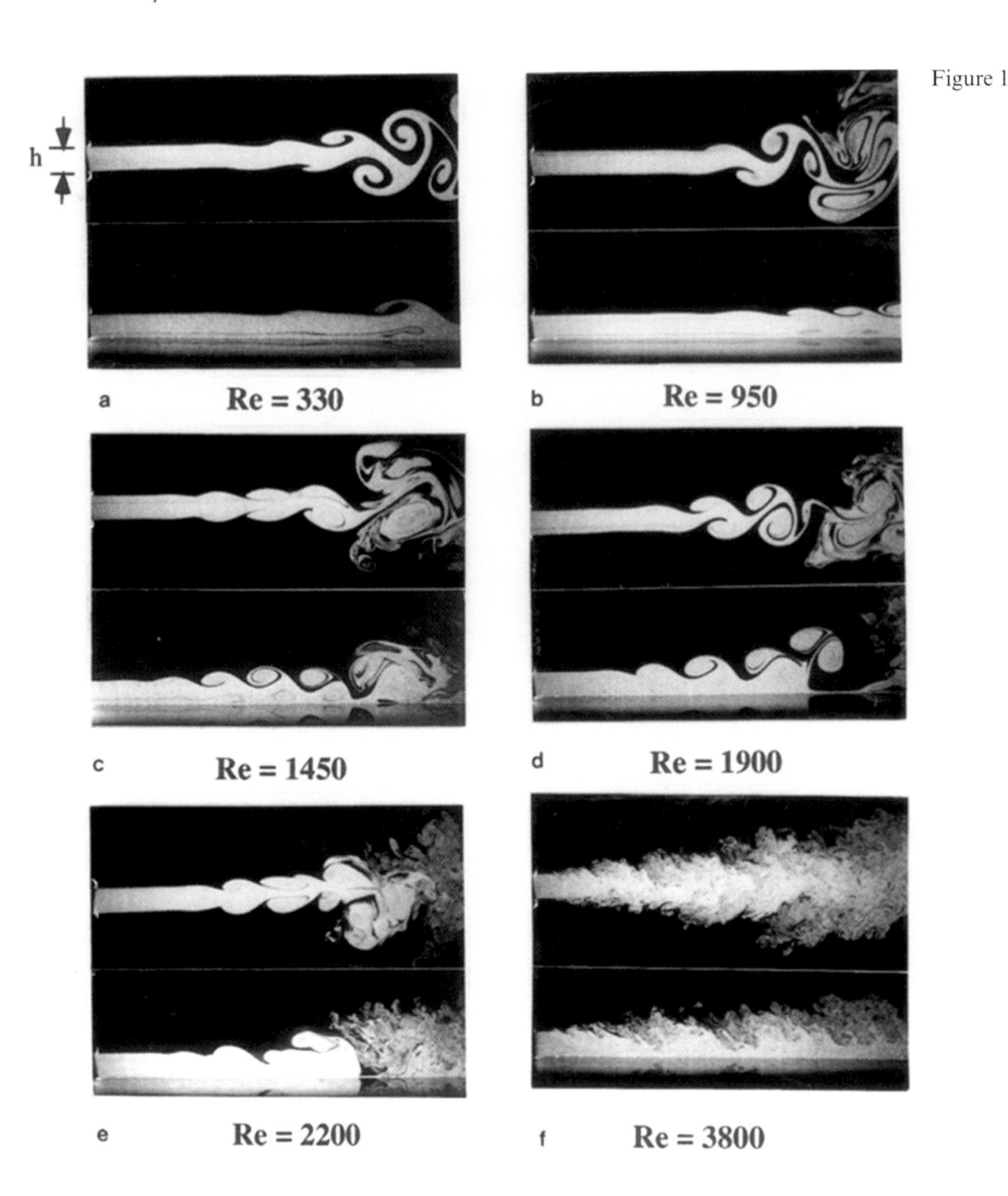

Figure 1

Comparison of transitional free jet and wall jet

S. Gogineni, C. Shih, and A. Krothapalli

FAMU/FSU College of Engineering, Florida A&M University and Florida State University, Tallahassee, FL 32310

The above flow pictures, extending from the jet exit to 15 widths downstream location for the range of exit Reynolds numbers (based on average exit velocity and width of the channel) from 300 to 3,800 were obtained using laser sheet/smoke flow visualization. These jets were issued from a rectangular channel (aspect ratio 20:1) with exit velocity profile being parabolic. At a low Reynolds number range (Figs. [a] and [b]), the growth rate of the wall jet is suppressed compared with the corresponding free jet. In other cases (Figs. [c] – Fig. [f]), no appreciable difference in growth rate is observed. In the case of wall jet, a double row vortex structure can be seen clearly (e.g., Fig. [c]), and during the transition process ejection of the boundary layer from the wall is noticeable (Fig. [d]). In addition, the flow evolution from laminar to turbulent state with increasing Reynolds number can be seen from Figs. (a) to (f) for both free jets and wall jets.

Keywords

smoke visualization; boundary-layer ejection; vortex pairing.

2 Vortices

Periodic axisymmetric vortex breakdown in a cylinder with a rotating end wall

J. M. Lopez and A. D. Perry

Aeronautical Research Laboratory, Melbourne, Australia

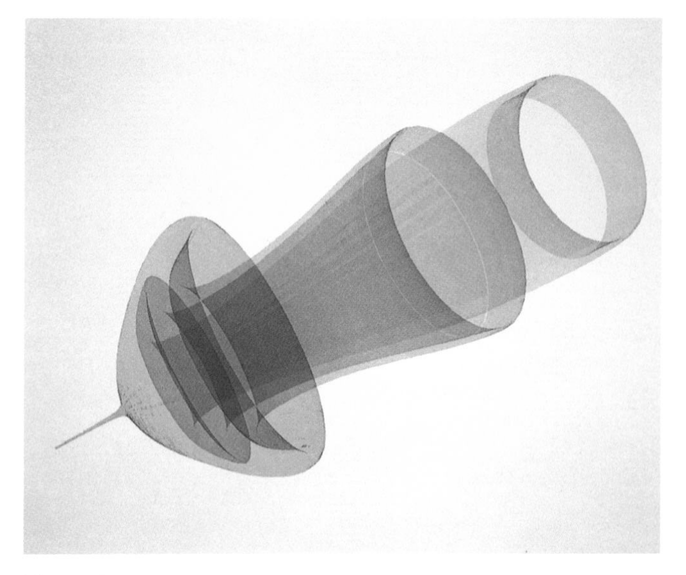

Figure 1

When the fluid inside a completely filled cylinder is set in motion by the rotation of the bottom end wall, steady and unsteady axisymmetric vortex breakdown is possible.[1,2] The onset of unsteadiness is via a Hopf bifurcation.

Figure 1 is a perspective view of the flow inside the cylinder where marker particles have been released from an elliptic ring concentric with the axis of symmetry near the top end wall. This periodic flow corresponds to a Reynolds number Re = 2765 and cylinder aspect ratio H/R = 2.5. Neighboring particles have been grouped to define a sheet of marker *fluid* and the local transparency of the sheet has been made proportional to its local stretching. The resultant *dye sheet* takes on an asymmetric shape, even though the flow is axisymmetric, due to the unsteadiness and the asymmetric release of marker particles. When the release is symmetric, as in Fig. 2, the dye sheet is also symmetric. These two figures are *snapshots* of the dye sheet after three periods of the oscillation (a period is approximately 36.3 rotations of the end wall). Figure 3 is a cross section of the dye sheet in Fig. 2 after 26 periods of the oscillation. Here only the marker particles are shown. They are colored according to their time of release, the oldest being blue, through green and yellow, and the most recently released being red. Comparison with Escudier's experiment[3] shows very close agreement.

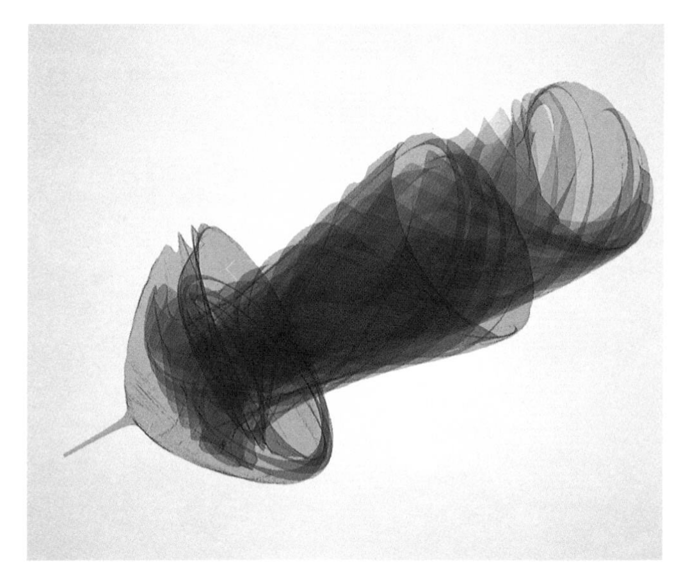

Figure 2

The particle equations of motion correspond to a Hamiltonian dynamical system and an appropriate. Poincaré map may be defined. Streaklines, such as in Fig. 3, can be used to approximate the unstable manifolds of the fixed points of the Poincaré map and these, together with the stable manifolds, can be used to describe in detail the filling and emptying of periodic axisymmetric vortex breakdown bubbles.[4]

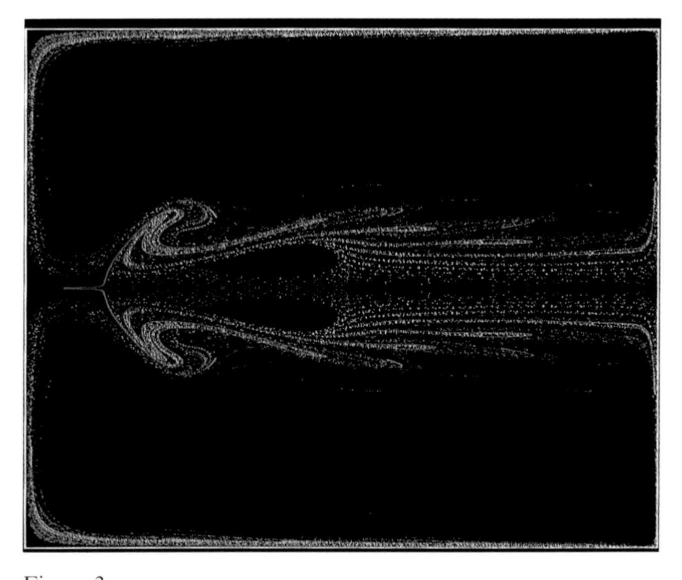

Figure 3

Keywords

simulation; Poincaré map; unstable manifold.

[1] J. M. Lopez, *J. Fluid Mech.* **221**, 533 (1990).

[2] G. L. Brown and J. M. Lopez, *J. Fluid Mech.* **221**, 553 (1990).

[3] M. P. Escudier, *Phys. Fluids* **28**, 2632 (1985).

[4] J. M. Lopez and A. D. Perry, *J. Fluid Mech.* **234**, 449 (1992).

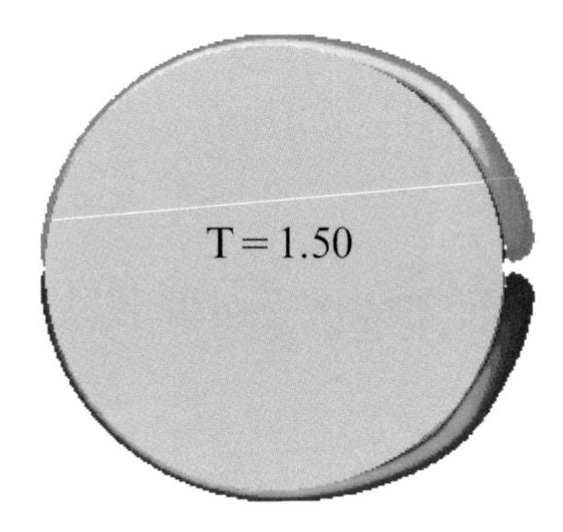

Figure 1

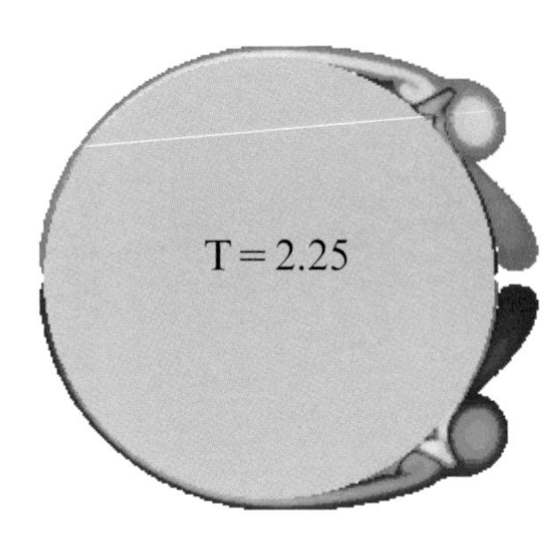

Figure 2

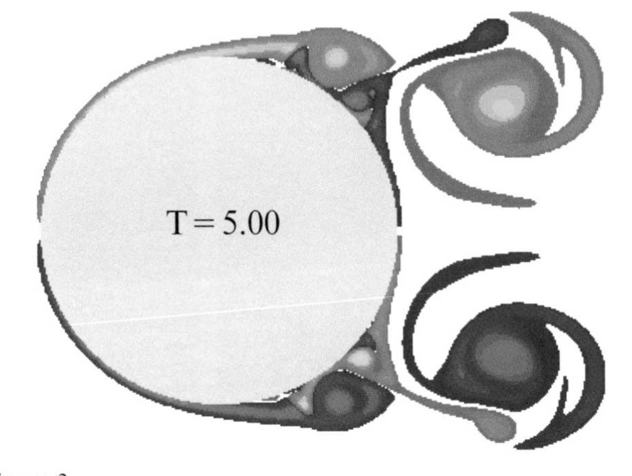

Figure 3

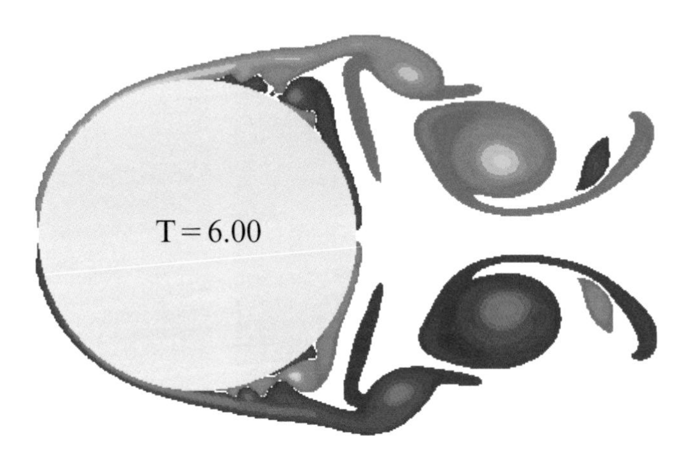

Figure 4

Flow past an impulsively started cylinder

P. Koumoutsakos and A. Leonard

California Institute of Technology

We conducted high resolution simulations of incompressible flow past an impulsively started cylinder at Re $=2UR/\nu=9500$. We solve numerically the vorticity form of the Navier-Stokes equations using a novel implementation of vortex methods, including an accurate treatment of diffusion effects and the no-slip condition.[1] At the onset of the impulsive start a very thin layer of primary (negative on the upper part of the cylinder) vorticity resides around the cylinder as it is introduced by the initial potential flow. This layer thickens and flows inducing a layer of secondary (positive on the upper half of the domain) vorticity on the backside of the cylinder. At $Ut/R=T\approx 1.5$ a bulge of vorticity forms at $\theta \approx 50°$ that is the signature of the unsteady separation. The secondary vorticity penetrates the feeding vortex sheet of the primary vortex at $T \approx 2.25$, and reaches the outer flow. The primary vortex then rolls up and detaches eventually from the body carrying along a part of the secondary vortex. The separating shear layer continues to extract vorticity from the surface of the body and a regular pattern of vorticity extraction, penetration of the secondary vorticity to the outer layer, and subsequent suppression by the oncoming shear layer is established. This and vortex merging in the near wake results in the intricate vorticity patterns shown at later times.

[1] P. Koumoutsakos and A. Leonard, "High-resolution simulations of the flow around an impulsively started cylinder using vortex methods," *J. Fluid Mech.* **296**, 1–38 (1995).

Keywords
simulation; start-up vortex; vortex shedding.

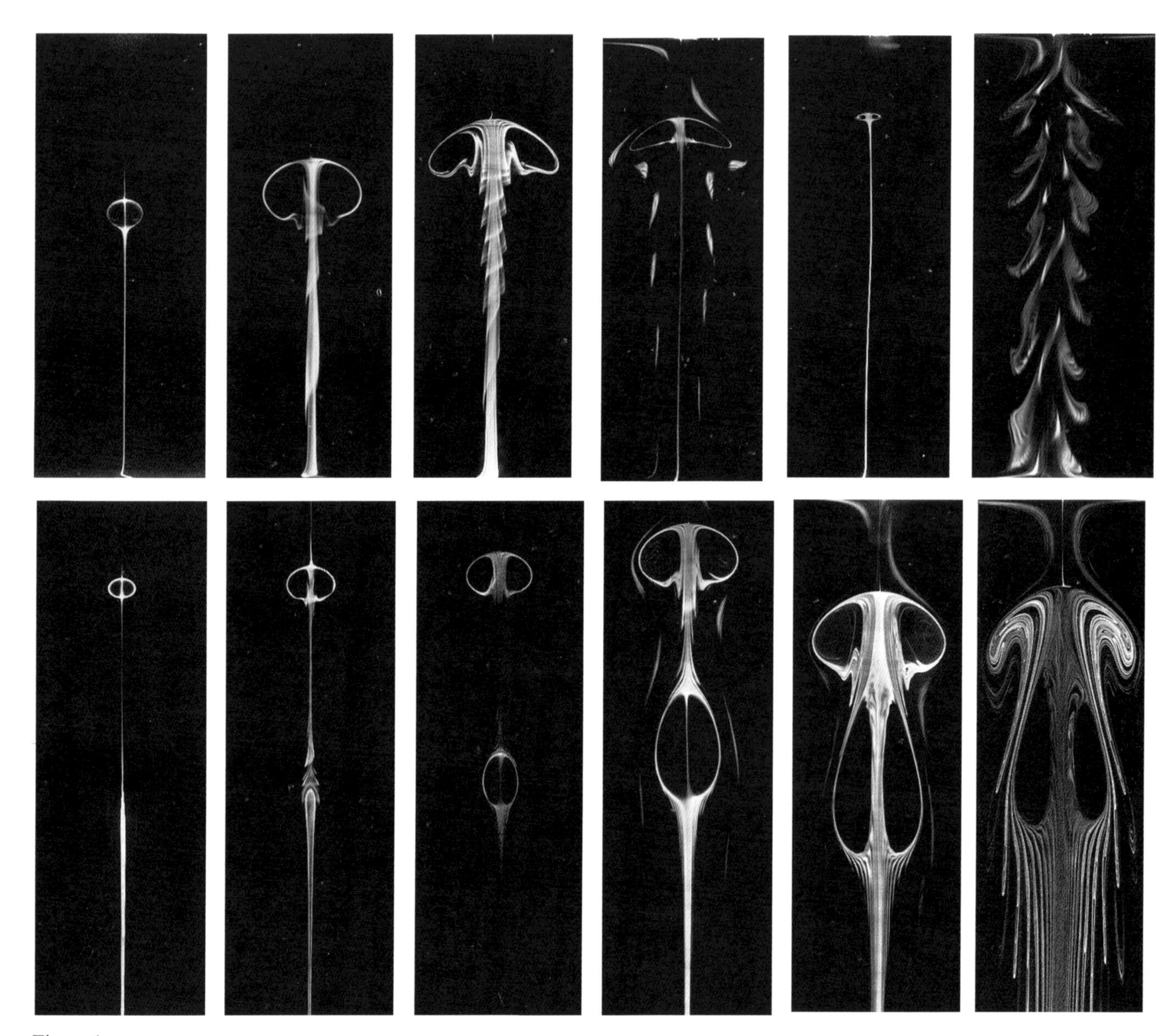

Figure 1

Swirling flow in a cylindrical container

M. P. Escudier

Brown, Boveri & Cie., Baden, Switzerland

This figure shows changes in vortex structure with increasing $\Omega R^2/\nu$ for $H/R = 1.75$ (upper row) and 2.5 (lower row). (By permission of Springer-Verlag, New York).

We produced the flow in a closed cylindrical container completely full of a glycerine–water mixture by rotating one endwall. We made a diametral plane visible by using laser-induced fluorescence. The two sequences of photographs above show the structure of the central core of the flow (30% of the total diameter). For a given value of the height-to-radius H/R (> 1.2), there are upper and lower limits to the rotation Reynolds number $\Omega R^2/\nu$ between which one, two, or three vortex breakdown bubbles occur on the axis of symmetry. For $\Omega R^2/\nu > 2500$ the flow becomes increasingly oscillatory and eventually turbulent. A complete description of the experiment has been given elsewhere.[1]

Keywords

laser-induced fluorescence; vortex breakdown; vortex bubble.

[1] M. P. Escudier, *Exp. Fluids* **2**, 189 (1984).

Figure 1(a)

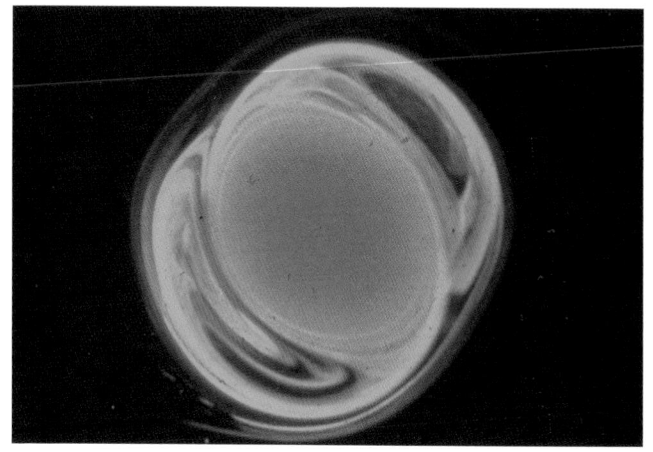

Figure 1(b)

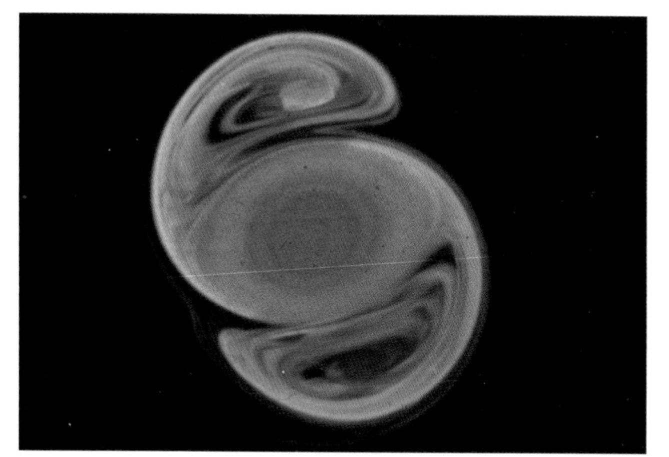

Figure 1(c)

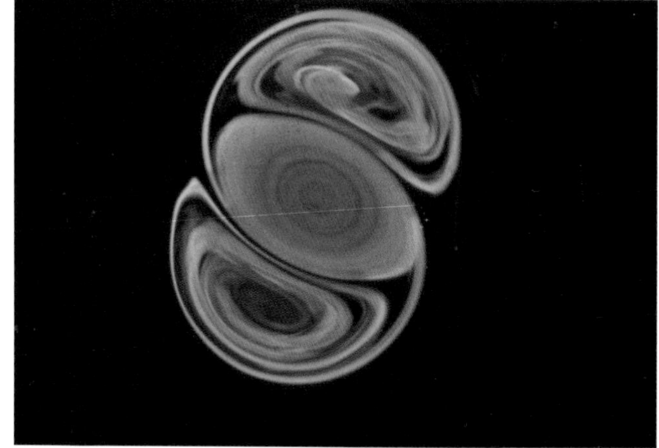

Figure 1(d)

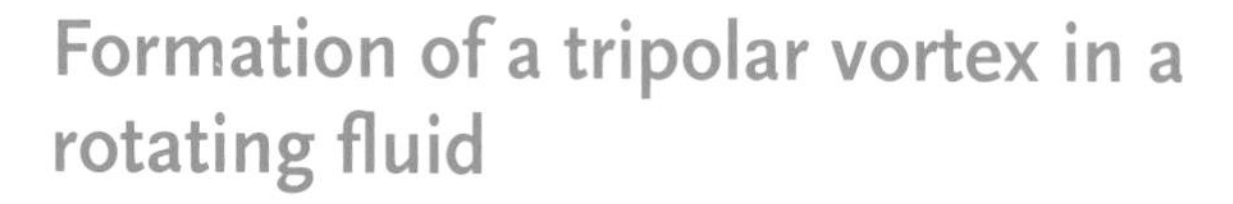

Formation of a tripolar vortex in a rotating fluid

G. J. F. van Heijst, R. C. Kloosterziel, and C. W. M. Williams

University of Utrecht/Eindhoven University of Technology, The Netherlands

The photographs displayed above show the formation of a tripolar vortex from an unstable axisymmetric vortex in a solidly rotating homogeneous fluid. The tripole is a symmetric, linear arrangement of three patches of distributed vorticity of alternate signs, and the axis of the vortex configuration rotates steadily about the center of the core vortex. In the experiment the initial vortex is an isolated cyclonic vortex, consisting of a core of cyclonic relative vorticity surrounded by a ring of anticyclonic relative vorticity. Under certain conditions such a vorticity distribution is unstable to perturbations of azimuthal wave number m. The photographs show an experiment in which $m=2$ is the most unstable mode, and the flow structure is observed to show a gradual transition into a stable tripolar vortex.

The tripole formation process is visualized by adding dye to the fluid. In the experiment shown here two different dyes were used in order to demonstrate that the tripole's satellite vortices result from a pure "rolling-up" of the outer ring of anticyclonic vorticity, whereas the vortex core (containing the cyclonic vorticity) is hardly affected by the instability, its shape only changing from circular to elliptical.

The (plan view) photographs were taken by a corotating camera at times $t=3.9T$ (a), $7.5T$ (b), $9.4T$ (c), and $12.8T$ (d), with $T=6.4$ sec the rotation period of the system, and $t=0$ corresponding with the release of the initial axisymmetric vortex. Further experimental details are described elsewhere.[1]

Keywords

vortex generation; vortex instability; dye visualization.

[1] G. J. F. van Heijst, R. C. Kloosterziel, and C. W. M. Williams, *J. Fluid Mech.* **225**, 301 (1991).

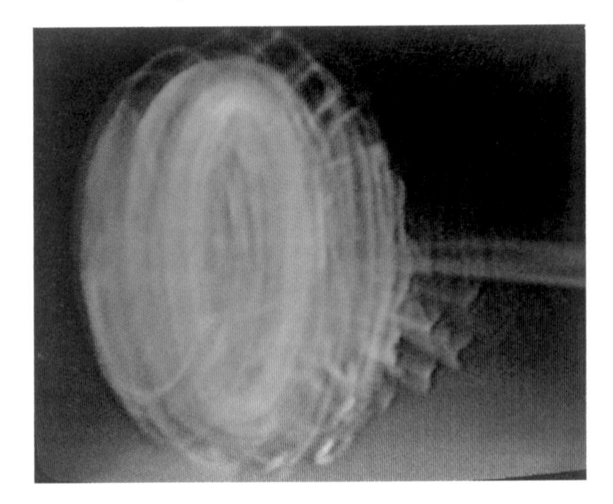

Figure 1 $T = 2.78$.

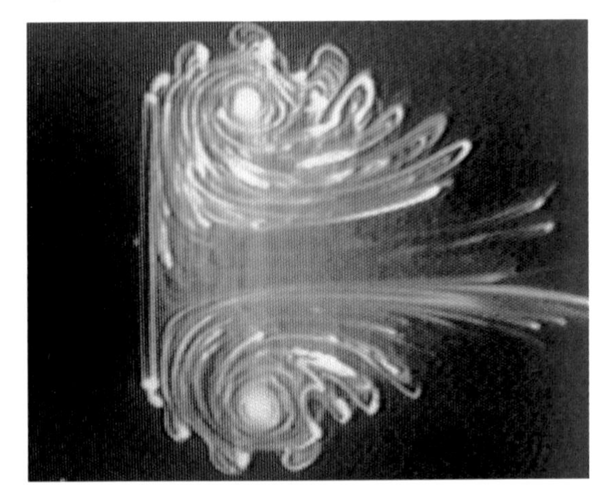

Figure 2 $T = 3.07$.

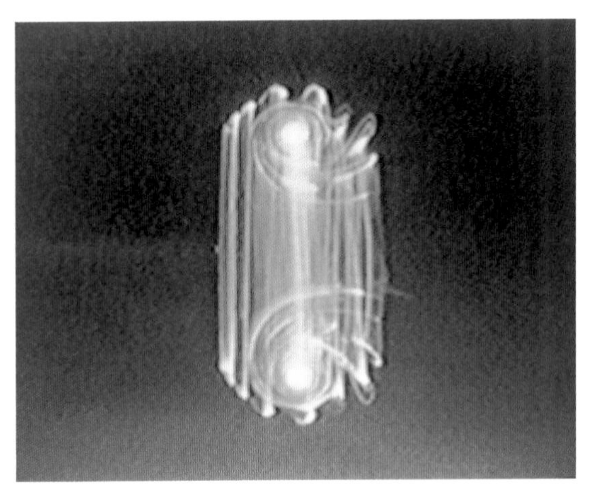

Figure 3(a) $T = 1.87$.

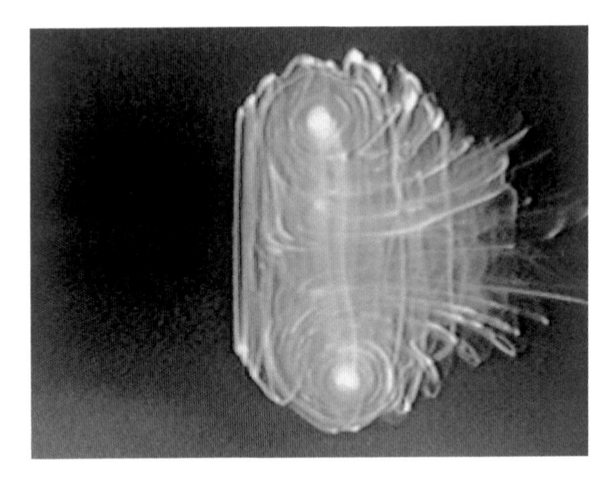

Figure 3(b) $T = 3.95$.

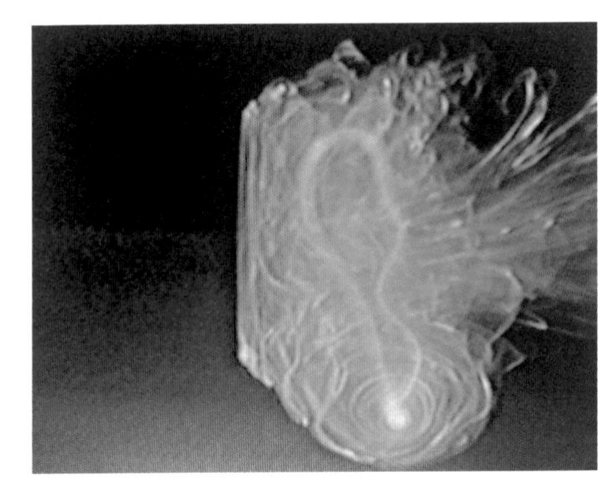

Figure 3(c) $T = 5.87$.

Vortex structure behind a disk started from rest

Hiroshi Higuchi and Hélène Balligand

Syracuse University

This work investigates the development and breakdown of the vortex structure behind a circular disk that is nearly impulsively started and subsequently moved at a constant speed.[1] Figure 1 shows an oblique view of the axisymmetric vortex that develops initially: this visualization is obtained by fluorescent dye and a UV light. In Fig. 2 the wake is illuminated only on its cross section by a laser light sheet where axisymmetric vortex roll-up and Helmholtz shear layer instability waves can be seen. With continued motion of the disk at the constant velocity, the vortex ring starts to tilt as the rear stagnation point moves off-center, as seen in a sequence of photographs [Figs. 3(a)–3(c)]. Here, a laser light sheet and a UV light are used simultaneously. The onset of three-dimensional structure is seen in Fig. 3(c). In Figs. 2 and 3, the acceleration rate, $A_p = D(dU/dt)U^2_0$, is 3 and the Reynolds numbers based on model diameter, Re_D, are 6200 and 4200, respectively ($A_p = 0.74$, $\mathrm{Re}_D = 9300$ for Fig. 1). Nondimensional time, $T = U_0 t/D$, is indicated in each figure.

[1] H. Higuchi, H. Balligand and J. H. Strickland, "Numerical and Experimental Investigations of the Flow Over a Disk Undergoing Unsteady Motion," *J. of Fluids and Structures* **10**, 705–719 (1996).

Keywords

fluorescent dye; UV-light; laser sheet; start-up vortex.

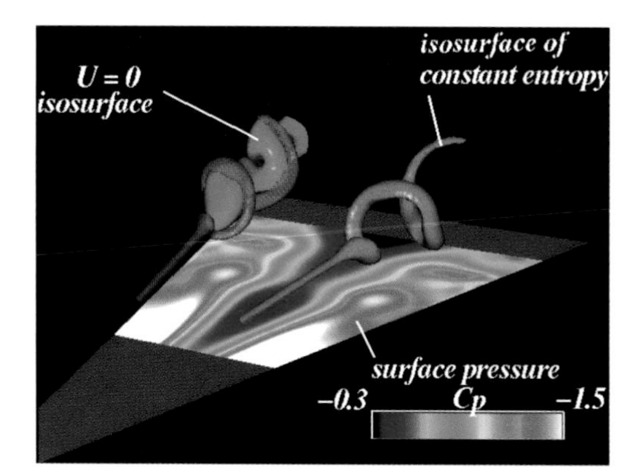

Figure 1

Figure 2

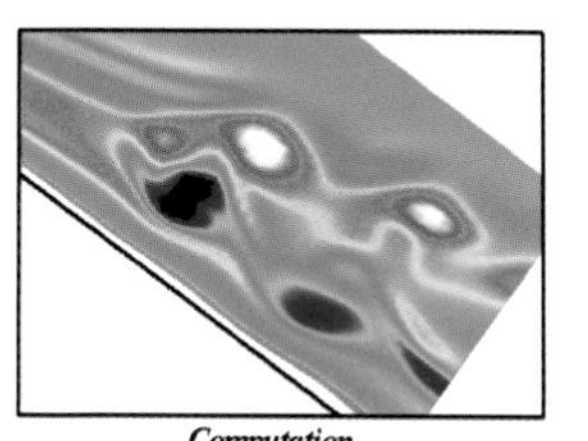

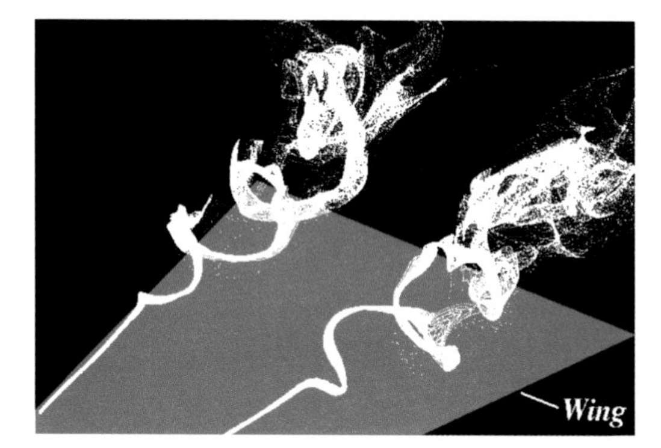

Figure 3

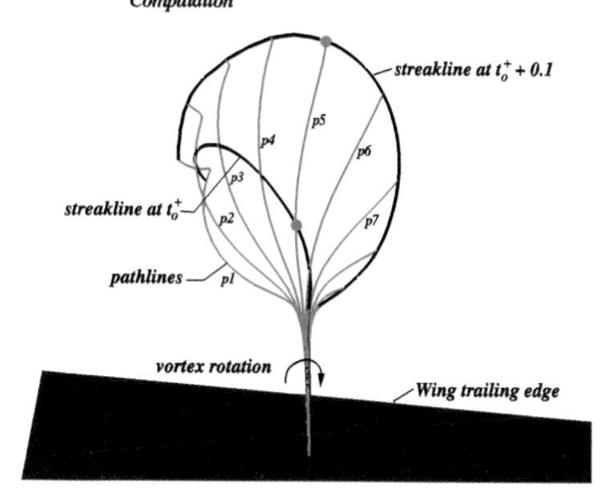

Figure 4

Computed unsteady structure of spiral vortex breakdown on a delta wing

M. Visbal

Air Force Research Laboratory
Wright–Patterson AFB, Ohio

The unsteady structure of spiral vortex breakdown above a 75° sweep delta wing was investigated numerically through direct solution of the full unsteady compressible Navier–Stokes equations.[1] The visualizations presented above correspond to an angle of attack of 34° and a chord Reynolds number of 9200. The 3-D instantaneous representation of Fig. 1 (taken from the video animation) clearly shows the sudden transformation of the rectilinear vortex core (depicted in red) into a whirling spiral of winding sense opposite to the swirling flow. A region of axial flow reversal (depicted in green) is contained within the spiral winding, and its outer envelope and internal structure are found to fluctuate dramatically in time. A strong interaction between the spiral breakdown and the wing surface also occurs as evidenced by the surface pressure contours. The instantaneous spanwise vorticity distribution on a plane normal to the wing and passing through the vortex center (Fig. 2) is characterized by staggered concentrations of vorticity at the locations where the spiral pierces the plane. These concentrations appear to propagate downstream as the spiral rotates. The computed unsteady structure was found to be in qualitative agreement with available PIV measurements.[2]

For the purpose of comparison with experimental observations, the numerical equivalent of streakline visualizations (Fig. 3) were generated by releasing material particles near the vortex axis upstream of breakdown. Computed pathlines and streaklines in the nose of the breakdown region are shown in Fig. 4. The streakline forms a spiral with winding opposite to the vortex and it rotates in time with the same sense of the upstream swirl. The material particles, however, do not follow spiral paths but instead are deflected radially away from the axis and proceed initially over the flow reversal region with little rotation consistent with conservation of angular momentum. Although the particles decelerate sharply as they approach breakdown, a true 3-D stagnation point in the instantaneous velocity field does not exist for spiral breakdown in front of the region of axial flow reversal.

Keywords

simulation; pathlines; streaklines; surface pressure.

[1] M. Visbal, "Computed unsteady structure of spiral vortex breakdown on delta wings," *AIAA* Paper No. 96–2074, 27th *AIAA* Fluid Dynamics Conference, June 1996.

[2] C. Shih and Z. Ding, "Unsteady structure of leading-edge vortex flow over a delta wing," *AIAA* Paper No. 96–0664, 34th Aerospace Sciences Meeting, January 1996.

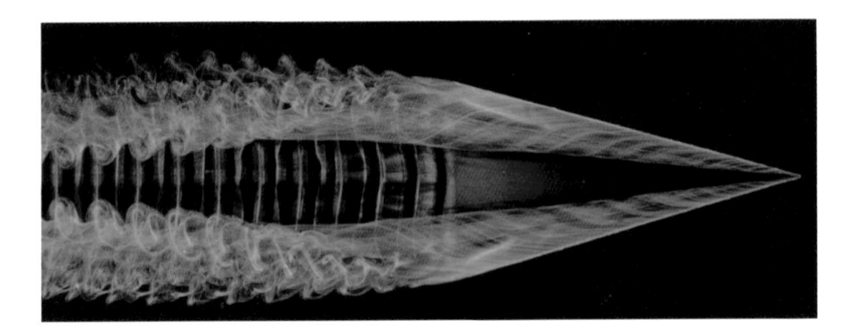

Figure 1

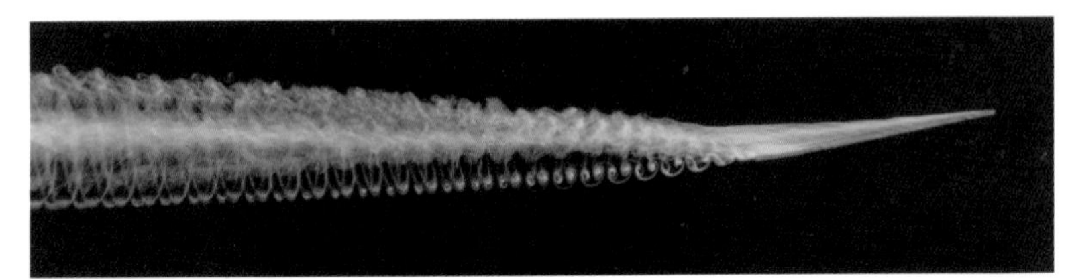

Figure 2

Figure 3

Free flight of a delta wing

G. D. Miller and C. H. K. Williamson
Cornell University

A great many studies have been directed towards understanding the flow over wings and, in particular, over a delta wing. Very surprisingly indeed, in our view, there exist *no* visualizations, from the laboratory, of the *far-field* development of the trailing vortex pair as it travels downstream, to our knowledge. Our flow visualization, involving novel free-flight gliding of a delta wing in water, shows for the first time, the exquisitely beautiful structure of the turbulent wake, in plan view (Fig. 1) and in side view (Fig. 2). One should note that these photographs are all to closely the same scale! The fluorescence dye, illuminated by a laser, shows that the near wake comprises an interaction between the primary streamwise vortex pair with the "braid" wake vortices between the pair.[1,2] Far downstream (64 chordlengths behind the wing) the primary vortex pair have reconnected and become large-scale rings (Fig. 3), although with a distinctly smaller length scale than predicted from Crow's analysis.[3]

This work is supported by the Office of Naval Research.

Keywords
turbulent wake; vortex pairs; vortex braids; laser-induced fluorescence.

[1] G. D. Miller and C. H. K. Williamson, "Trailing vortex instabilities in the wake of a delta wing," *Bull. Am. Phys. Soc.* **39**, 1878 (1994).
[2] G. D. Miller and C. H. K. Williamson, "Turbulence in the wake of a delta wing," in *Advances in Turbulence VI*, (ed. S. Gavrilakis, L. Machiels and P. A. Monkewitz), Kluwer, 187 (1996).
[3] S. C. Crowe, "Stability theory for a pair of trailing vortices," *AIAA J.* **8**, 2172 (1970).

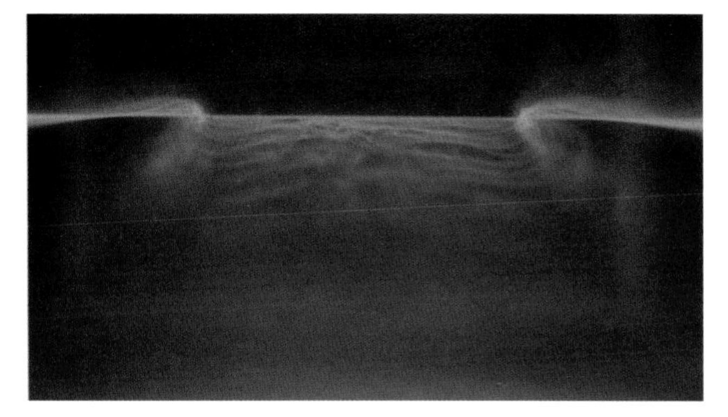

Figure 1 Model study (end view).

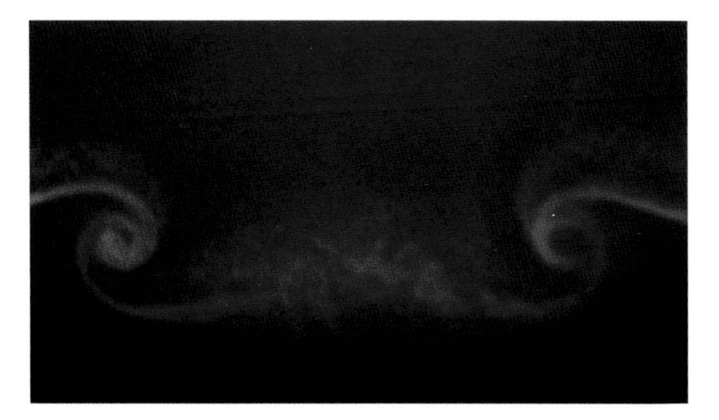

Figure 2 Model study (cross-sectional view).

Figure 3 Aircraft wake (photo courtesy of Cessna Aircraft Company).

Aircraft trailing vortices and downwash phenomenon

Hiroshi Higuchi

Syracuse University

A simple water channel experiment shows the phenomenon of tip vortices and downwash similar to that behind an aircraft despite large differences in scaling parameters. In this study, a rectangular flat plate of 5 cm chord length with an aspect ratio of 4.2 was placed at 4.5° angle of attack in the water channel with a 0.6 m×0.6 m×2.44 m test section. The Reynolds number based on the chord length was 5.7×10^3.

The hydrogen bubble technique was used for flow visualization. The photograph in Fig. 1 was taken from downstream with a thin platinum wire located slightly below the model and 18 chord lengths downstream. In Fig. 2, a laser sheet placed at 28 chord lengths downstream of the trailing edge illuminated the cross-sectional view of the vortices. The effect of the separated flow at the trailing edge is also visible near the midspan.

The trailing vortices and downwash phenomenon of an aircraft in flight are seen clearly in Fig. 3. In this situation, a Cessna Citation VI was flown immediately above the fog bank over Lake Tahoe at approximately 313 km/h or 170 knots.[1] Aircraft altitude was about 122 m (400 ft) above the lake, and the weight was approximately 8400 kg. As the trailing vortices descended over the fog layer due to the downwash, the flow field in the wake was made visible by the distortion of the fog layer. The photo was taken by P. Bowen for the Cessna Aircraft Company from the tail gunner's position in a B-25 flying in formation slightly above and ahead of the Cessna. The aircraft is seen initiating a gentle climb after a level flight, leaving a portion of the fog layer yet unaffected. The wing span measured 16.3 m and the wing area was 29 m^2. The Reynolds number based on the mean aerodynamic chord of 2.1 m was 1.1×10^7. Results from other experimental and computational studies[2,3] also compare favorably with flow structures shown here.

[1] B. Budzowski, Director of Flight Operations, Cessna Aircraft Company (private communication, 1993).

[2] J. R. Spreiter and A. H. Sacks, "The rolling rep of the trailing vortex sheet and its effect on the downwash behind wings," *J. Aeronaut Sci.*, January, 21 (1951).

[3] R. Krasny, "Computation of vortex sheet roll-up in the Trefftz plane," *J. Fluid Mech.* **184**, 123 (1987).

Keywords

downwash far-field; Cessna aircraft.

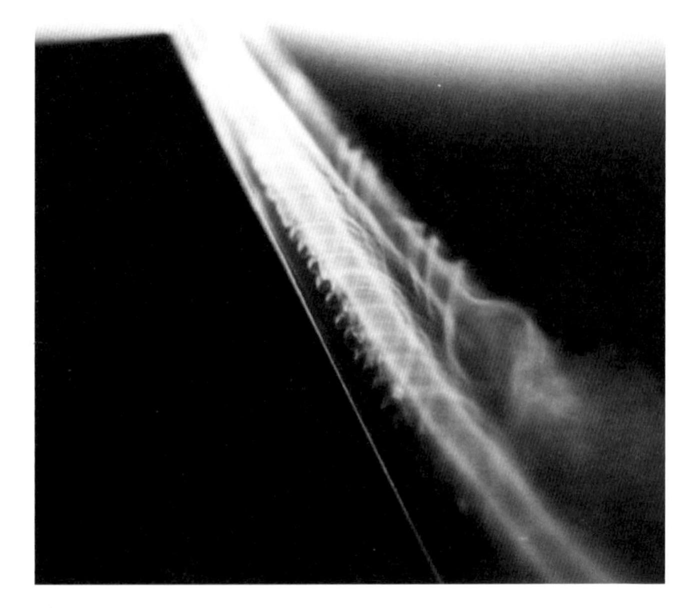

Figure 1

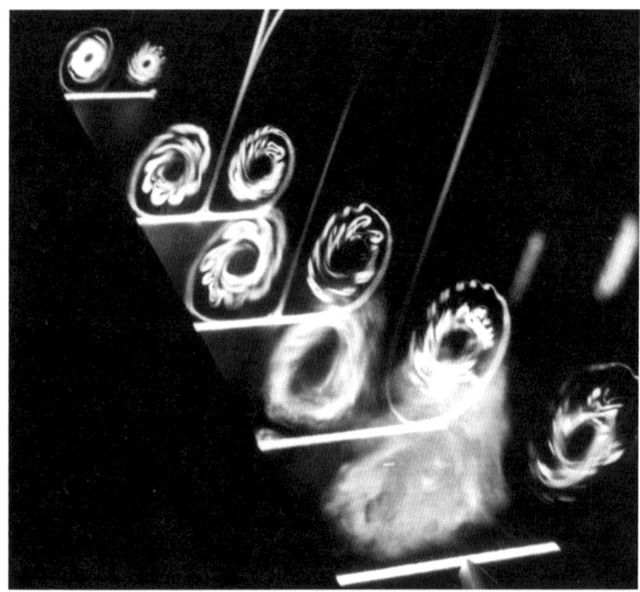

Figure 2

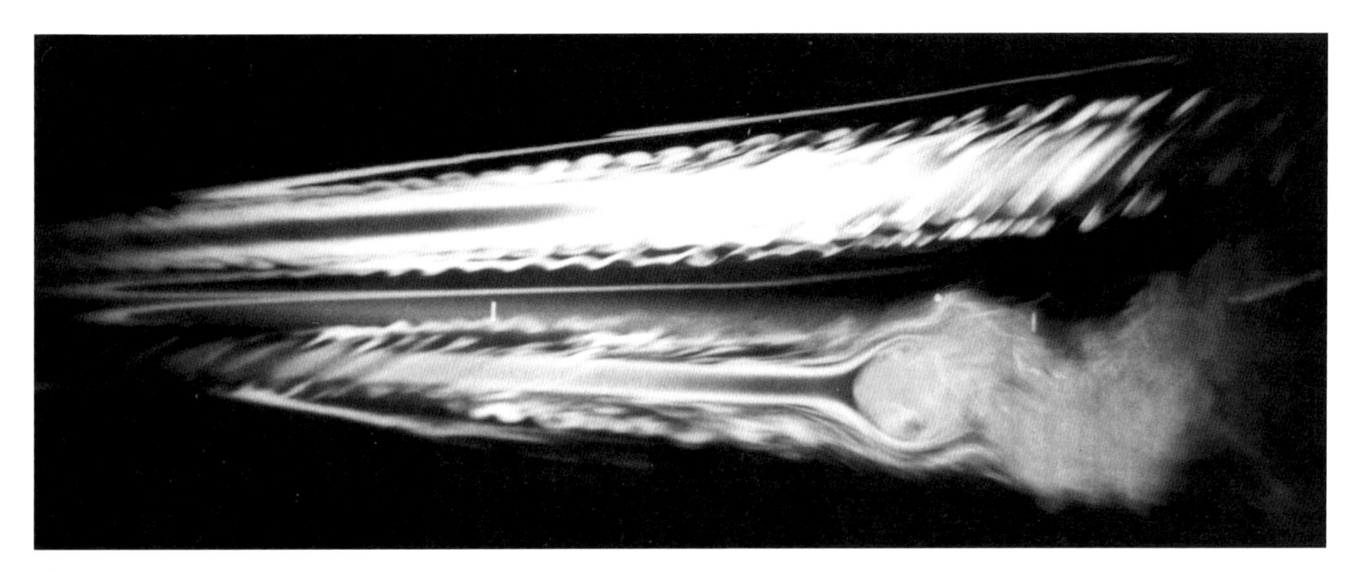

Figure 3

Vortex breakdown on a delta wing

F. M. Payne, R. C. Nelson, and T. T. Ng
University of Notre Dame

The above photographs show the breakdown of a leading-edge vortex on an 85° swept delta wing at a 40° angle of attack. The freestream velocity is 10 ft/sec. For highly swept wings at high angles of attack, the vortices tend to break down asymmetrically. In Fig. 1, a column of smoke introduced upstream of the test section impinges on the apex of the model and is entrained into the vortices. The right vortex is breaking down at $x/c = 0.50$. Figure 2 was obtained using the laser sheet technique. It is a multiple exposure showing five lateral cross sections of the flow field. The fourth cross section from the apex is cutting directly through the "bubble" region of the breakdown. Figure 3 is a longitudinal cross section of the vortices obtained by rotating the laser sheet by 90°. A region of high-speed axial flow occurs in the core of each vortex and appears in the photographs as an area void of smoke.

Keywords
smoke visualization; laser sheet; vortex bubble.

Figure 1

Interaction of two-dimensional wakes

Q. Rahaman, A. Alvarez-Toledo, B. Parker, and C. M. Ho

University of Southern California

Multiple closely spaced bluff bodies are moved through a thin soap film illuminated by monochromatic light. A *truly* two-dimensional wake flow is produced and indicated by the deformed diffraction patterns of soap film. The picture shows the wake of three cylinders placed in line and an equal distance apart with a ratio between the surface of the cylinders and cylinder diameter of 5.7. The towing speed of the bodies was about 35 cm/sec. with the flow from left to right.

Keywords
light illumination; soap film; bluff-body wake.

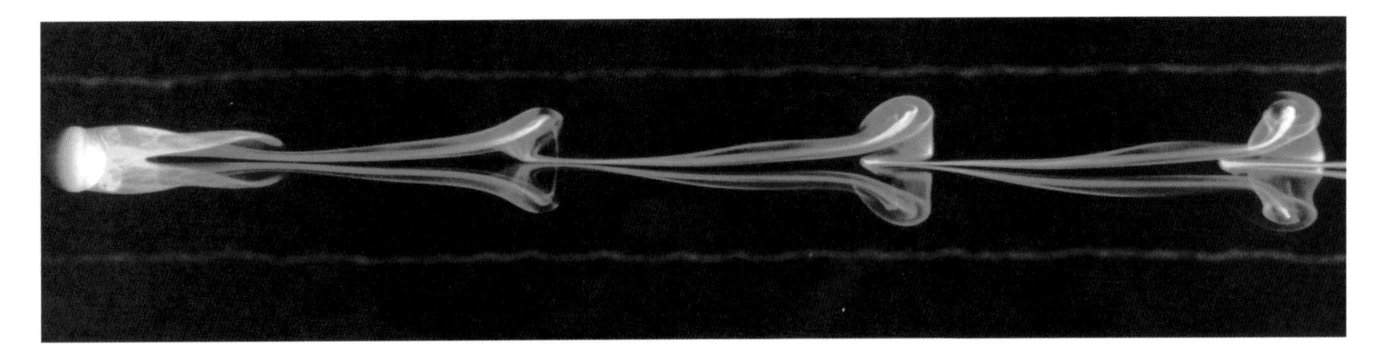

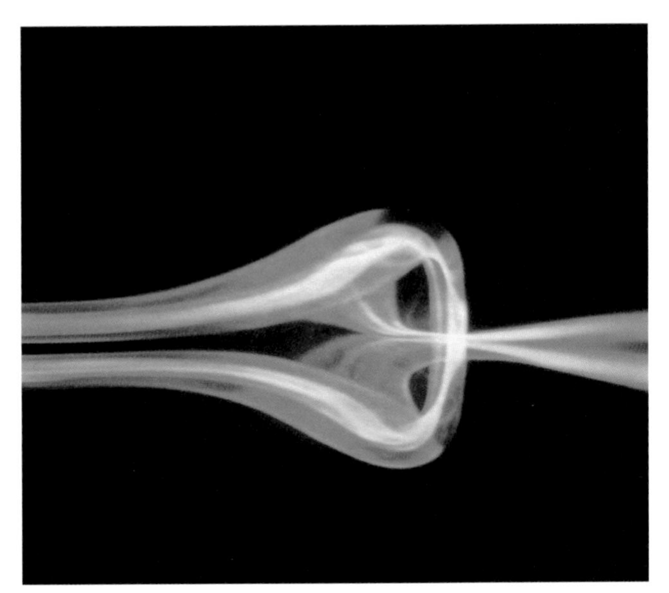

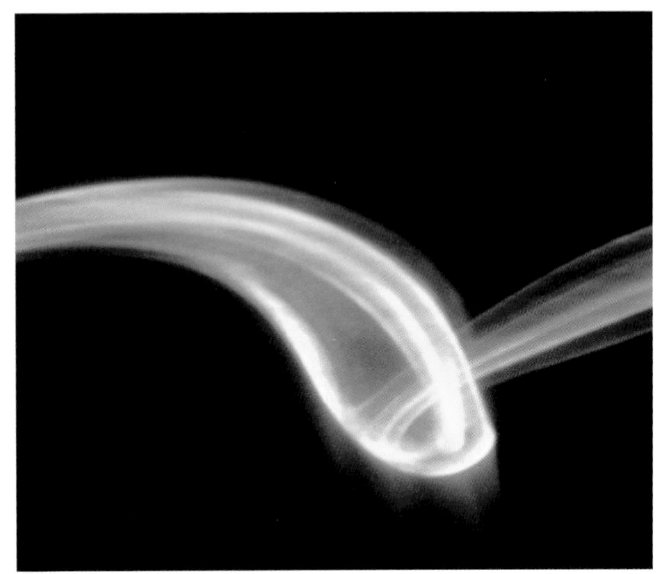

Figure 1

Vortex dynamics in the wake of a sphere

T. Leweke, M. Provansal, D. Ormières, and R. Lebescond

IRPHE, CNRS/Universités Aix-Marseille, France

We present experimental visualizations of the flow structure in the wake of a sphere, at a Reynolds number just above the threshold for the onset of periodic wake oscillations. The flow was studied in a water channel, where the sphere of diameter 1 cm was held by a small rod from upstream. Visualization was achieved using fluorescent dye illuminated by laser light.

The first transition occurring in the sphere wake is associated with a loss of the wake's rotational symmetry, while the flow still remains stationary. In this state the recirculation region immediately behind the sphere divides itself into two parallel threads.

The upper photograph shows the flow at a Reynolds number of 320, based on the sphere diameter. Here the wake has undergone a second transition to a time-dependent state, which is characterized by a periodic shedding of vortex loops, connected by counterrotating pairs of vortex filaments. When the wake is viewed from a different angle, it may be (and has often been) misinterpreted as having a helical structure. However, the upper image clearly shows the existence of a reflectional symmetry with respect to a plane going through the sphere center.

The lower photographs show a close-up of the oscillating wake at about seven diameters behind the sphere, seen simultaneously from two perpendicular directions. They give a detailed view of the connection between the horseshoe-type vortex loop and the trailing legs of the loop shed in the previous cycle of the wake oscillations.

These visualizations demonstrate that, even at very low Reynolds numbers, the wake of a sphere placed in a uniform flow has a surprisingly complex spatial structure.

Keywords

laser-induced fluorescence; bluff-body wake; vortex loop.

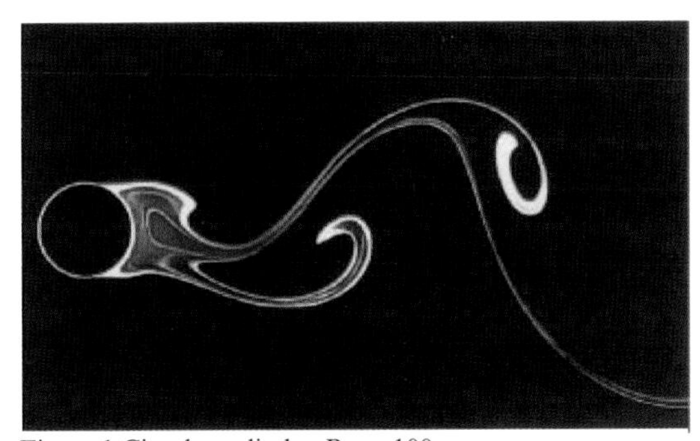

Figure 1 Circular cylinder, Re = 100.

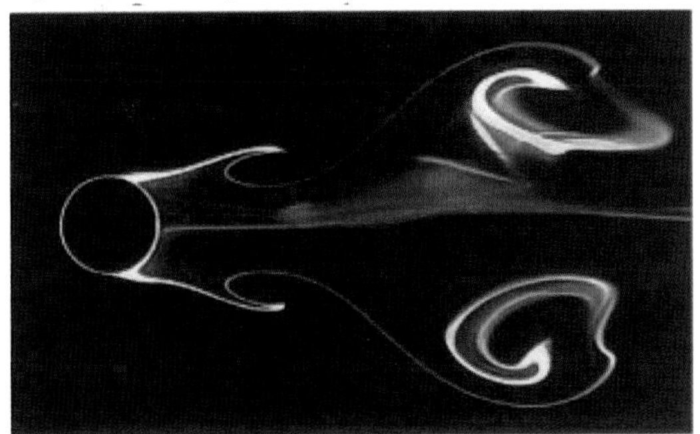

Figure 2 Wide wake.

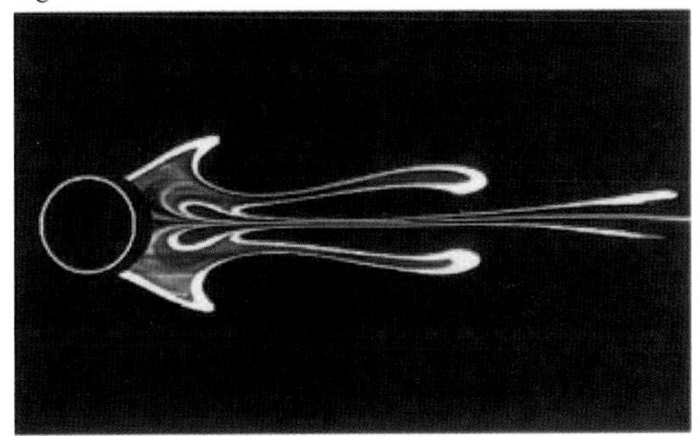

Figure 3 Narrow wake.

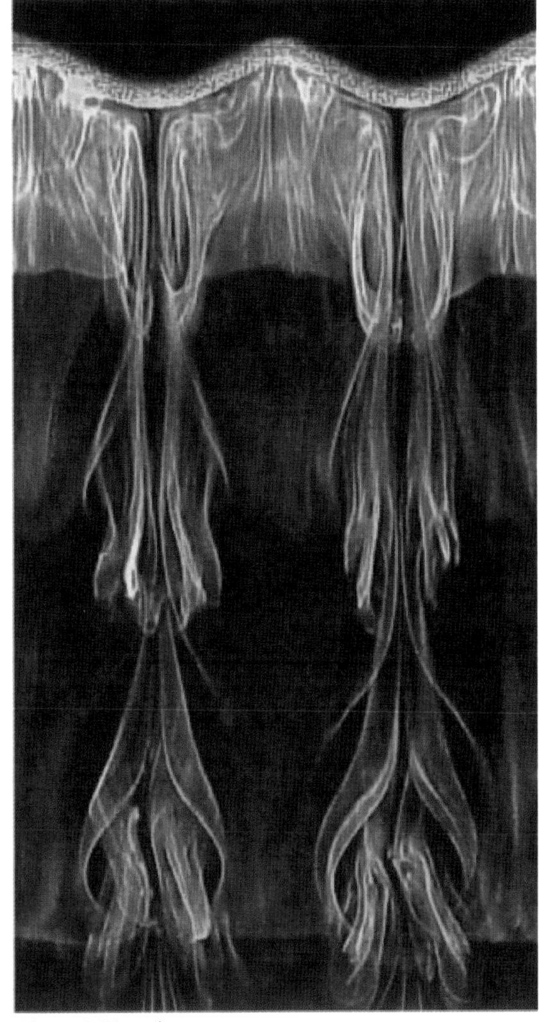

Figure 4 Plan view.

Suppression of Kármán vortex shedding

J. C. Owen *Imperial College*

A. A. Szewczyk *University of Notre Dame*

P. W. Bearman *Imperial College*

The photographs show flow visualization results from towing tank experiments to study flow past a sinuous bluff body. The body has a circular cross section with constant diameter D along the span and an axis that is sinuous. All the visualizations are for a Reynolds number, based on D, of 100. For the model shown, the wavelength is equal to $7.5D$ and the peak-to-peak wave height is equal to $0.166D$. These experiments were carried out in the Hydraulics Laboratory of the Department of Aeronautics at Imperial College and the flow was visualized using laser-induced fluorescence.

Key features of the results are that von Kármán type vortex shedding (Fig. 1) is suppressed and there is observed to be a periodic variation in the wake width across the span. A wide wake (Fig. 2) is found where the body protrudes farthest downstream at a trough and a narrow wake (Fig. 3) is found behind a peak. The plan view (Fig. 4) shows the flow to be quite different to that behind a cylindrical bluff body, with vortex loops trailing behind the troughs. The results shown here are in agreement with earlier work on rectangular sections with a wavy front face[1] where shedding was suppressed and substantial drag reductions were recorded.

J.C.O. acknowledges the support of EPSRC and A.A.S. the support of ONR via Grant No. N00014-96-1-0756.

Keywords

laser-induced fluorescence; tow tank; vortex shedding.

[1] P. W. Bearman and J. C. Owen, "Reduction of bluff body drag and suppression of vortex shedding by the introduction of wavy separation lines," *J. Fluids Struct*. **12**, 123 (1998).

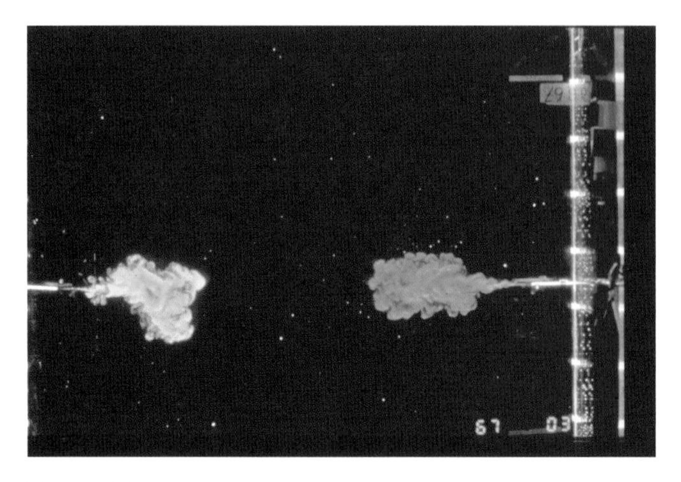

Figure 1(a)

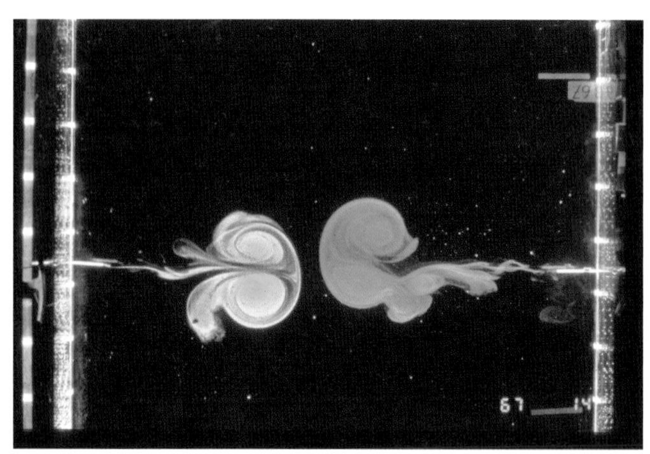

Figure 1b)

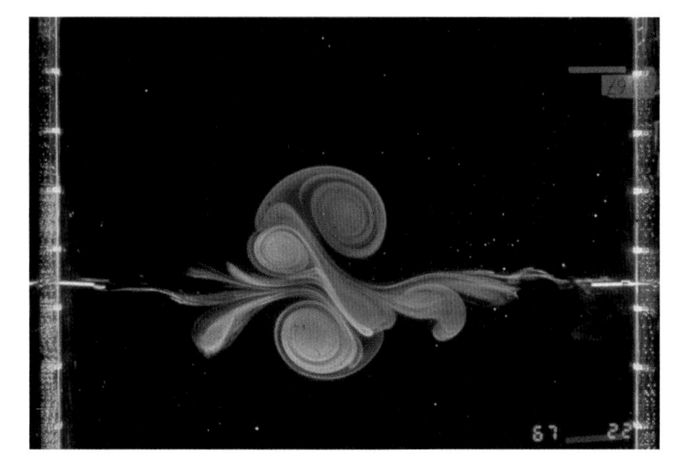

Figure 1(c)

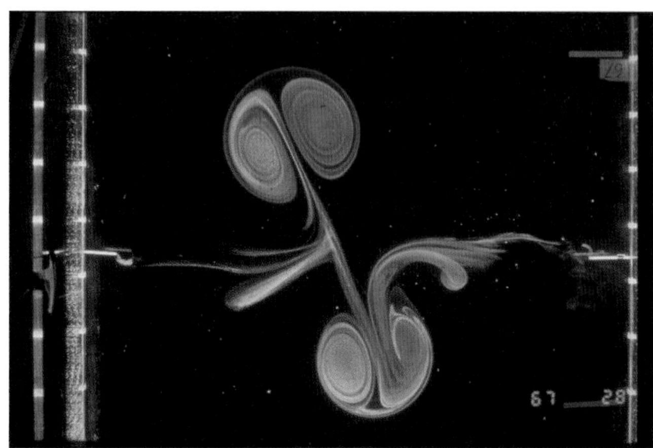

Figure 1(d)

Colliding dipolar vortices in a stratified fluid

G. J. F. van Heijst and J. B. Flór
University of Utrecht, The Netherlands

The photographs illustrate experiments that have been performed on the collapse of a three-dimensional turbulent patch in a linearly stratified fluid. The turbulence was generated by horizontal injection of a small volume of fluid during a short time interval. A transition to two-dimensional flow occurs when the turbulent patch collapses under gravity, as can be observed from the spectral flux of kinetic energy to larger scales. The collapsed fluid eventually gets organized in a dipolar flow structure that moves slowly forward along a straight line. The robustness of this dipolar coherent structure is demonstrated in experiments on head-on collisions of two dipoles with approximately identical characteristics. Consecutive stages of a head-on collision are shown by the (plan view) photographs, taken (a) 42 sec, (b) 70 sec, (c) 120 sec, and (d) 225 sec after the injections were stopped. The asymmetry in the observed flow patterns is due to a slight misalignment of the initial dipoles. Nevertheless, the experiment shows nicely that the original dipoles exchange partners, and that two new dipoles emerge, moving along straight lines away from the collision area. Further experimental details are described elsewhere.[1]

Keywords
liquid dye; vortex dipole; vortex collision.

[1] G. J. F. van Heijst and J. B. Flór, *Nature* **340**, 212 (1989).

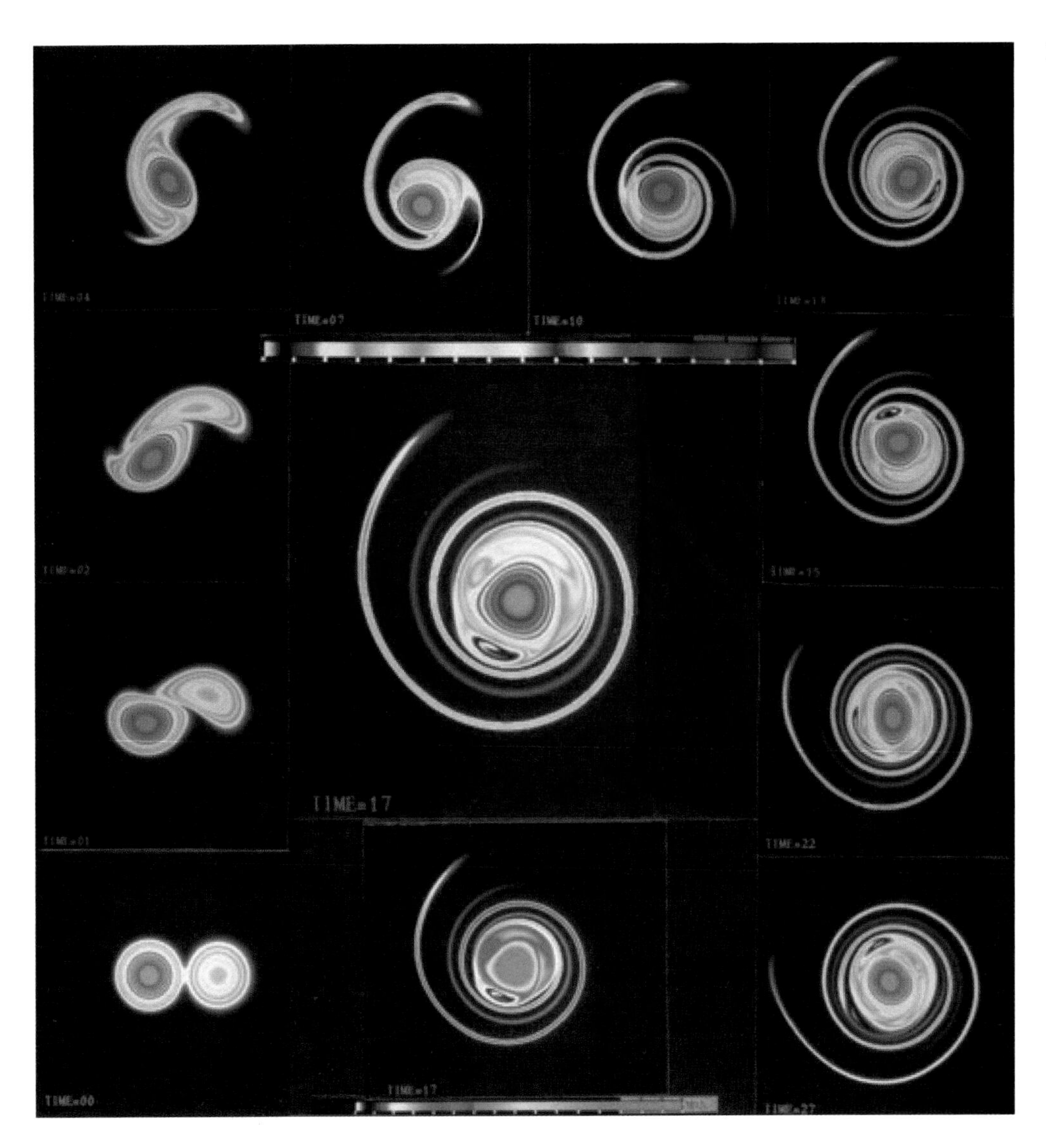

Figure 1

Asymmetric vortex merger

C. Seren, M. V. Melander, and N. J. Zabusky

University of Pittsburgh

The figure displays evolution of a fundamental interaction in two-dimensional incompressible turbulence. The visualization sequence shows the merger of vorticity obtained from a numerical simulation[1] using a pseudospectral algorithm on a (256^2) 2-D mesh. The large centrally placed figure at $t = 17$ shows a near threefold symmetric core region and the fine underlying structure in the low-vorticity regions. The centrally placed color map shows colors assigned in a linear fashion with the vorticity (from left to right). (Below this figure is the same function at $t = 17$ rendered with a different color map, which emphasizes low-lying levels.)

The simulation data were displayed on a color monitor, with a resolution of 480^2, 8 bit deep pixels. The pictures were taken with a 135 mm lens.

The simulations in this study were made on the CRAY-1 computer at the National Center for Atmospheric Research, which is supported by the National Science Foundation. The research work in 2-D vortex dynamics is supported by the Army Research Office and the Office Naval Research. The work station on which the images were created was provided by SUN Microsystems, Inc., with an additional grant by the National Science Foundation.

Keywords

simulation; vortex pairing.

[1] M. V. Melander, N. J. Zabusky, and J. C. McWilliams, *Phys. Fluids* **30**, 2610 (1987).

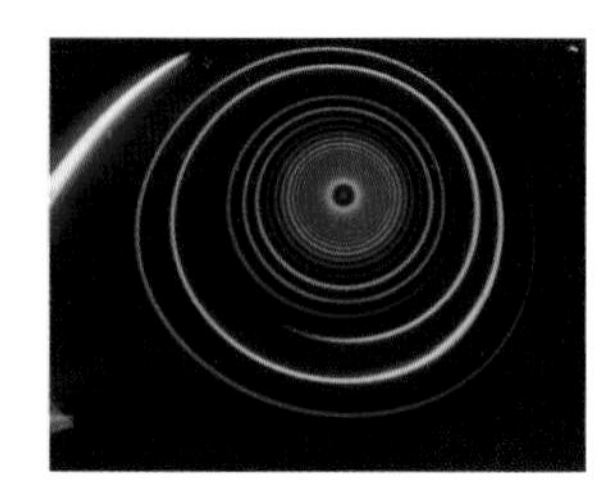
Figure 1

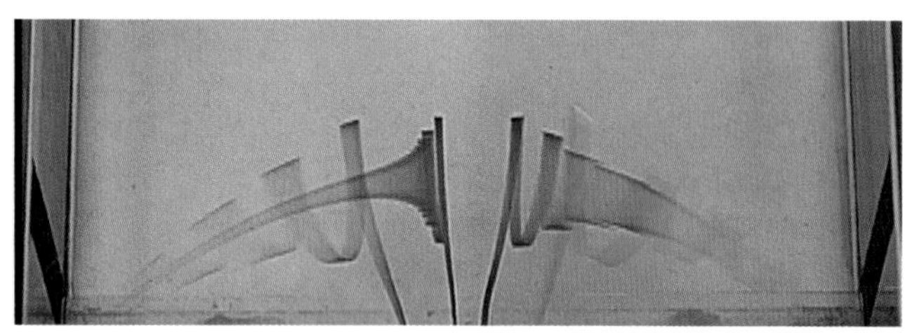
Figure 2

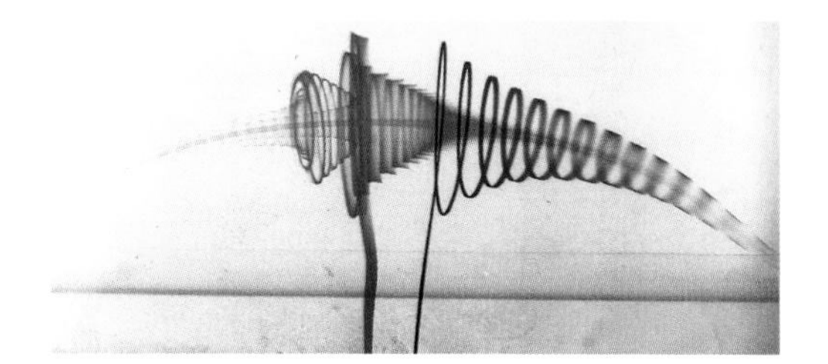
Figure 3

Figure 4

Figure 5

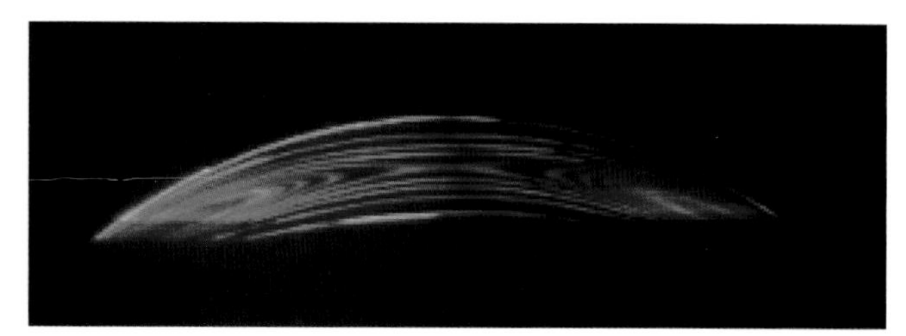

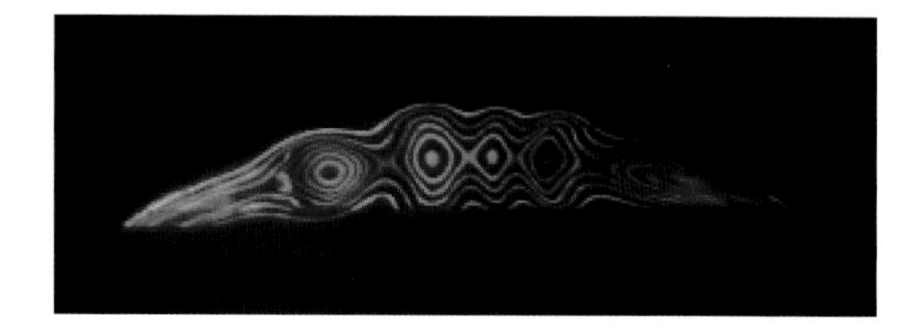

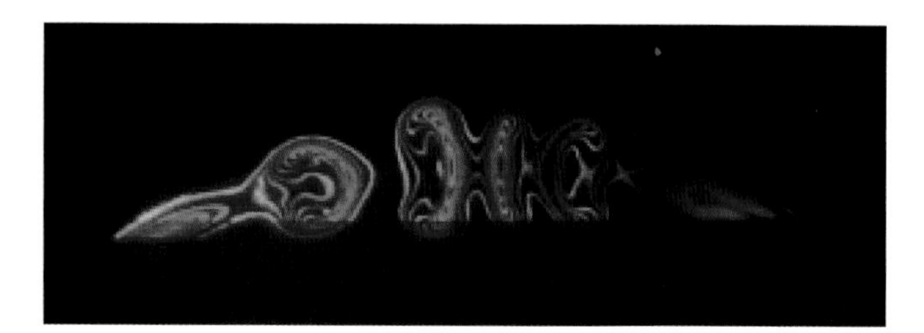
Figure 6

Visualizations of vortex filaments

Philippe Petitjeans

Laboratoire de Physique et Mécanique des Milieux Hétérogènes, ESPCI, Paris, France

Visualizations of stretched vortices are presented which model concentrated vorticity filaments in turbulent flows. These are known to play an important role in the intermittency. Because of the difficulties associated with working on filaments in real turbulence, we isolate such a structure from its turbulent background. A vortex filament is created by stretching (through suction) the vorticity of a laminar boundary layer flow. This coherent structure is visualized and characterized as a function of the initial vorticity and of the stretching.

A stable vortex can be created as visualized in Fig. 1 where a cross section of the spiral structure is observed, or by injecting dye jets from upstream (Figs. 2 to 5). Depending on the parameters (flow rate of the suction, and main flow rate of the channel), the vortex either persists or breaks up. In the latter case, the vortex is generated, then begins to oscillate, and breaks up into a turbulent spot. Even in the former case, the vortex can be unstable, and it produces pairs of counterrotating rolls around and along itself (Fig. 6). These rolls appear around the main vortex with a periodicity that depends linearly on the stretching. The instability is centrifugal in nature. The Rayleigh criterion on the azimuthal velocity $v_\theta(r)$ can be negative outside the core of the vortex.

Keywords

dye visualization; stretched vortex; centrifugal instability.

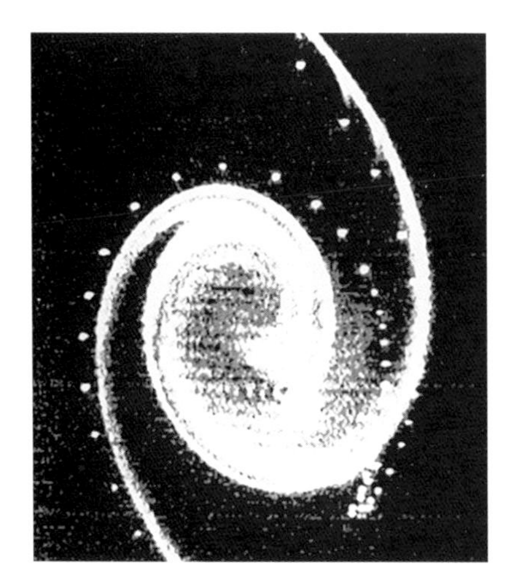

Figure 1 1 droplet/vortex.

Figure 2 Droplet entrainment.

Figure 3 10 droplets/vortex.

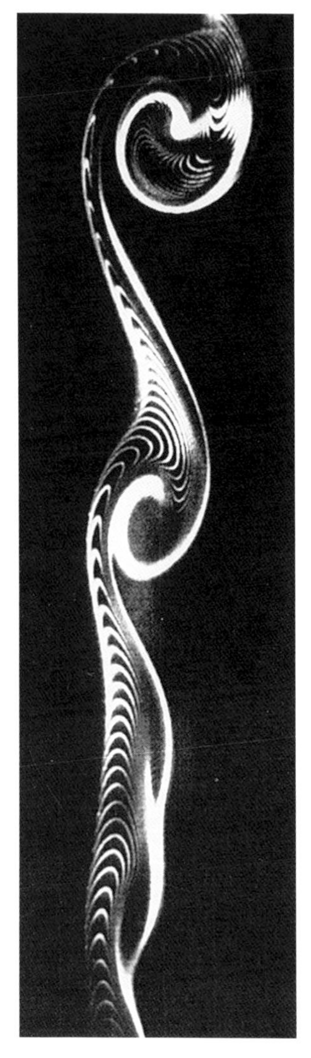

Figure 4 25 droplets/vortex.

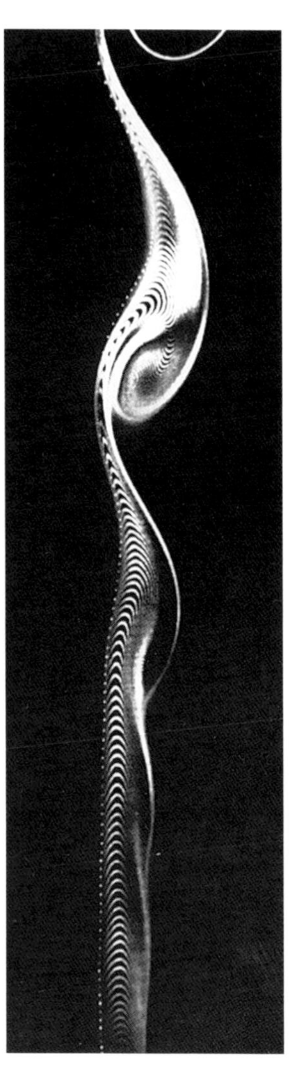

Figure 5 50 droplets/vortex.

Visualization of "Vaporlines" Emanating From Water Droplets

Robert Hancock

Air Force Research Laboratory
Wright–Patterson AFB, Ohio

These images illustrate a novel visualization technique in which water droplets, vapor from these droplets, and their interaction with the carrier gas can be observed simultaneously as they interact with one another in two-phase flows. Specifically, water droplets ($<$70 μm) are injected into a two-dimensional, $TiCl_4$-laden, gaseous flow where they evaporate. The water vapor reacts spontaneously with the $TiCl_4$ vapor to form micron-sized TiO_2 particles. The particles are small enough to rapidly accelerate to the velocity of the carrier gas but are too large to readily diffuse. Thus they are convected along the path that the water vapor follows as it leaves the droplets. The instantaneous locus of the TiO_2 particles is defined as a "vaporline," and is visualized using Mie scattering and laser sheet lighting. A vapor line is similar to a streakine, however, a vaporline originates from a moving point source in the flow, whereas, a streakline originates from a fixed point source.

The images shown were obtained with single pulses of a frequency-doubled Nd:YAG laser and an unintensified CCD camera. The air velocity is 1 m/sec and 0.5 m/sec on the left and right of the splinter plate, respectively. The flow was driven acoustically at 20 Hz. The convective vaporlines remain fairly thin and well defined over a long distance for the two-dimensional gaseous flows investigated. This occurs largely because these flows are laminar and well behaved. Images collected at a variety of droplet-to-vortex ratios are shown.

This work was supported by AFOSR.

Keywords
laminar mixing layer; Mie scattering; chemical reaction; acoustic forcing.

3 Patterns

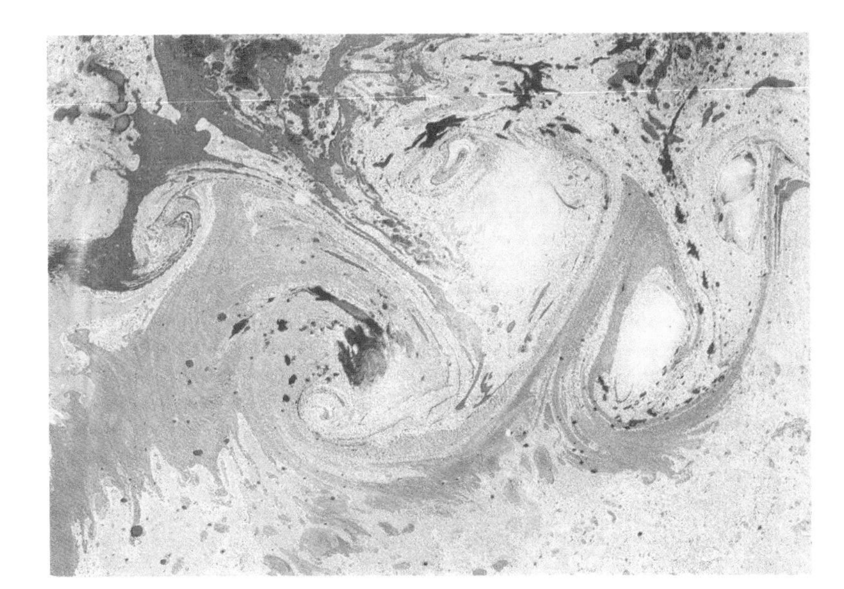

Figure 1

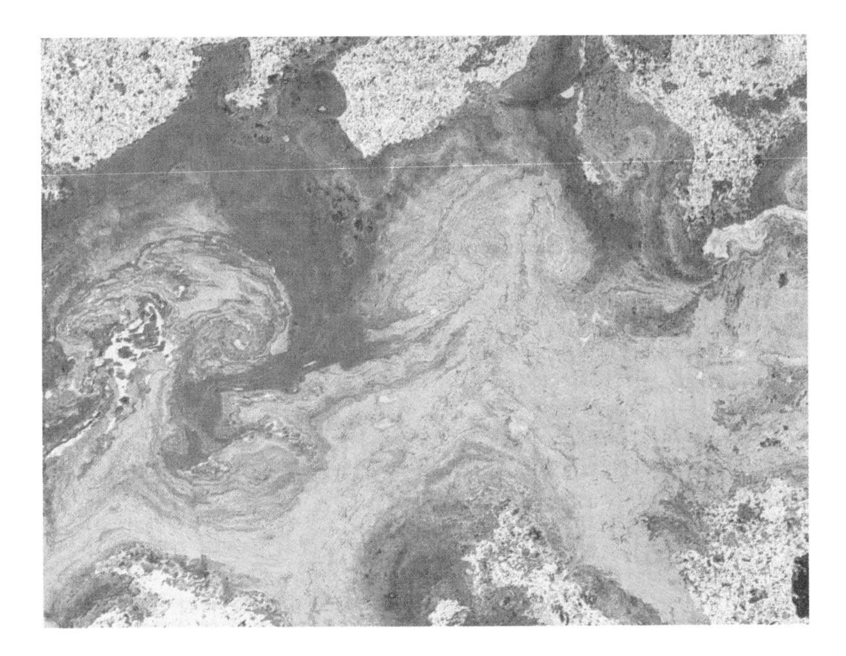

Figure 2

Vortex flows paint themselves

G. E. Koppenwallner and D. Etling
Universität of Hannover, Germany

The artistlike pictures of vortex flows presented here have been produced by the flow itself. The method of this "natural" flow visualization can be described briefly as follows: The working fluid is water mixed with some paste in order to increase the viscosity. Vortex flows are produced by pulling a stick or similar devices through the fluid or by injecting fluid through a nozzle into the working tank.

The flow visualization is performed in the following way: the surface of the fluid at rest is sparkled with oil paint of different colors diluted with some evaporating chemical. After the vortex structures have formed due to wakes or jets, a sheet of white paper is placed on the surface of the working fluid, where the oil color is attached to the paper immediately. The final results are artistlike paintings of vortex flows which exhibit a rich variety of flow structures.

Keywords
oil paint visualization; direct image-transfer.

Figure 1

Figure 2

Mixing in regular and chaotic flows

C.-W. Leong and J. M. Ottino

University of Massachusetts at Amherst

These photographs show the time evolution of two passive tracers in a low Reynolds number two-dimensional time-periodic flow. The initial condition corresponds to two blobs of dye, green and orange, located below the free surface of a cavity filled with glycerine. The flow is induced by moving the top and bottom walls of the cavity while the other two walls are fixed. In this experiment the top wall moves from left to right and the bottom wall moves from right to left; both velocities are of the form $U\sin^2(2\pi t/T)$, with the same U and the same period T, but with a phase shift of 90°.[1] The governing parameter is the wall displacement D; in this particular case the top and bottom walls move 3.9 times the width of the cavity in one period. The Reynolds number is 1.2 and the Strouhal number is 0.1. The tracer exhibits the stretching and folding of chaotic flows while the islands (regular regions) preclude widespread mixing. Note that the green and orange dyes do not mix; if D is increased to 4.42 they do. The top picture corresponds to 20 periods; the bottom picture to $20\frac{1}{4}$ periods. Note that the placement of the islands becomes symmetric with respect to the horizontal axis and the vertical axis. Blobs placed in islands barely stretch at all.

Keywords

passive tracer; time-periodic forcing; stretching and folding; fluorescent dye.

[1] C.-W. Leong and J. M. Ottino, "Experiments on mixing due to chaotic advection in a cavity," *J. Fluid Mech.* **209**, 463–499 (1989).

Figure 1 Mixing state of an initially segregated drop of ink in a two-dimensional stirring field.

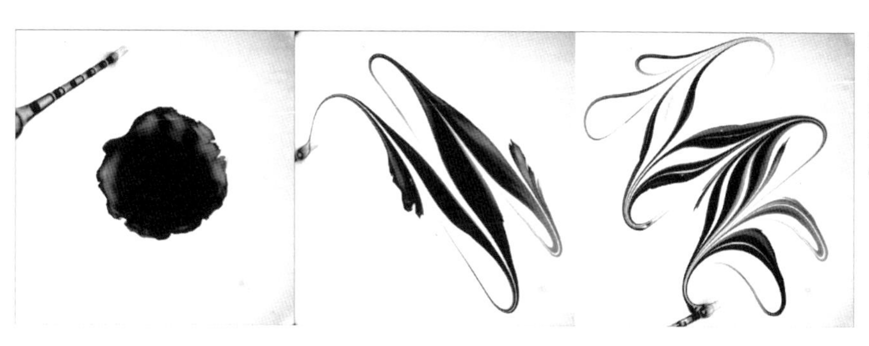

Figure 2 The stirring protocol of a drop of ink deposed at the surface of pure Glycerol using a small rod. The sequence displays the initial state, half, and a completed stirring cycle.

Mixture's route to uniformity by coalescence

E. Villermaux and J. Duplat
IRPHE and IUSTI, Marseille, France

The paradigm of the spoon stirring a drop of milk in a cup of coffee is often referred to when discussing mixing. We present the viscous version of it: the straw in a milk-shake. Stirring a blob of dye with a rod in a thin layer of a viscous fluid (Fig. 1) is instructive to understand how a mixture evolves towards uniformity.

The stirring protocol consists in slicing the medium in the plane of the fluid layer with a small rod. A number of parallel cuts is made in one direction, and then the same number at a right angle, this operation defining one cycle (Fig. 2).

At low Reynolds number (typically 10^{-1}) the medium is deformed by the rod on a scale given by its own size. The maximal rate of stretch is obtained for fluid particles close to the rod trajectory, while the protocol leaves nearly unstretched fluid parcels far from its trajectory. Those keep a concentration close to the initial concentration before being, possibly, stretched at the next cycle. Concomitantly, fluid particles are brought close to each other in the wake of the rod and coalesce. Coalescence results in the addition of the concentration levels of nearby particles and the protocol is such that this addition process is made at random. The mixture concentration distribution $P(C)$ thus evolves by *self convolution* and is actually found to be well described by a family of Gamma functions $P(C)=(n^n C^{n-1} e^{-n(C/\langle C\rangle)})/(\langle C\rangle^n \Gamma(n))$.

Since material lines grow in proportion to the number of cycles in this two-dimensional flow, the maximal concentration of one stretched, diffusing particle decays like (number of stirring cycles)$^{-[1+(1/2)]}$. The number of convolutions keeping the average concentration $\langle C\rangle$ constant cycle after cycle is thus $n \propto$(number of stirring cycles)$^{3/2}$, leading eventually to the mixture's uniformity $P(C, n\to\infty) \to (C-\langle C\rangle)$.

Keywords
liquid dye; chaotic mixing.

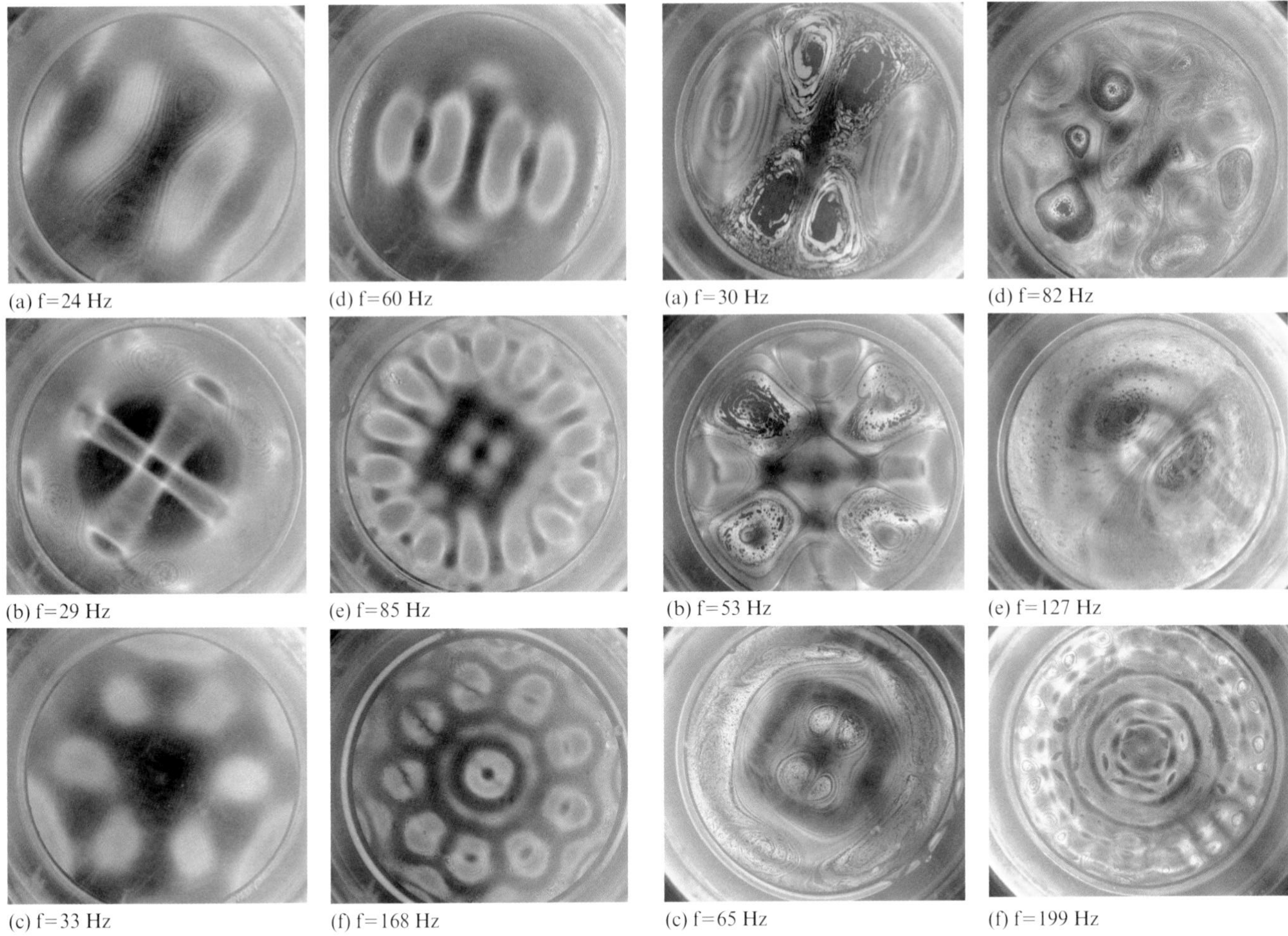

(a) f=24 Hz (d) f=60 Hz (b) f=29 Hz (e) f=85 Hz (c) f=33 Hz (f) f=168 Hz

Figure 1 Thick film shadowgraph images.

(a) f=30 Hz (d) f=82 Hz (b) f=53 Hz (e) f=127 Hz (c) f=65 Hz (f) f=199 Hz

Figure 2 Thin film interferences fringes.

The generation of two-dimensional vortices by transverse oscillation of a soap film

P. D. Weidman

University of Colorado, Boulder

V. O. Afenchenko, A. B. Ezersky, S. V. Kiyashko, and M. I. Rabinovich

Institute of Applied Physics, Russian Academy of Science, Nizhny Novgorod, Russia

Observations of the dynamics of horizontal soap films stretched over circular frames 8.0 cm in diameter undergoing periodic transverse oscillations at frequencies 20–200 Hz and accelerations 3–18 g are captured on 35 mm film. The composition was typically 94% tap water, 5% glycerine, and 1% liquid soap. Lighting from an overhead fluorescent lamp yields shadowgraph images of the flexural mode patterns when the film is relatively thick as in Fig. 1. As the film thins by evaporation, vortex motions are observed through colored fringe interference patterns as in Fig. 2.

Many of the mode patterns are spatially stationary. However, the pattern in 1(c) pulsated and the pattern in 1(d) oscillated as a whole chaotically, back and forth in the azimuthal direction. Vortices often appear in counter-rotating pairs around the perimeter of the cells, but also emerge spontaneously in the interior. As the film thins the vortices coalesce into larger vortices and observations suggest that the total circulation of the vortex system is zero. Inevitably, "black films" of thickness $O(10\text{ nm})$ emerge at the vortex centers as may be seen in Figures 2(a) and 2(e). Young vortices often lie between thick "bladders" of liquid as evidenced by the closely spaced fringe pattern in Figure 2(b) where we estimate the ratio of film thickness between the center of the bladder to the neighboring black films to be 1000:1 or more. More details of the experiment and a mathematical model that captures some of the observed vortex phenomena are given in references 1 and 2.

Keywords

shadowgraph; interferometry chaotic oscillations.

[1] V. O. Afenchenko *et al.*, *Phys. Fluids* **10**, 390 (1998).

[2] J. M. Vega *et al.*, *J. Fluid Mech.* **372**, 213 (1998).

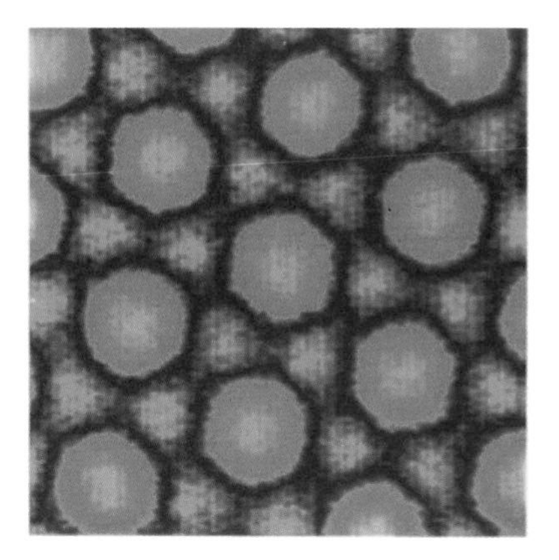

Figure 1 Hexagons magnified, $\epsilon = 0.007$.

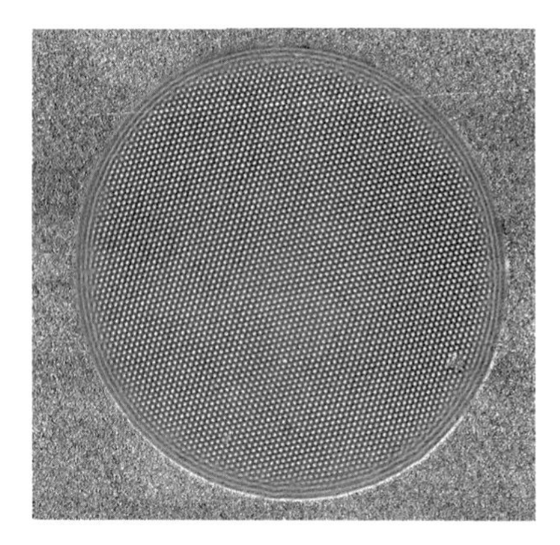

Figure 2 Hexagons, $\epsilon = 0.06$.

Figure 3 Two-armed rotating spiral, $\epsilon = 0.15$.

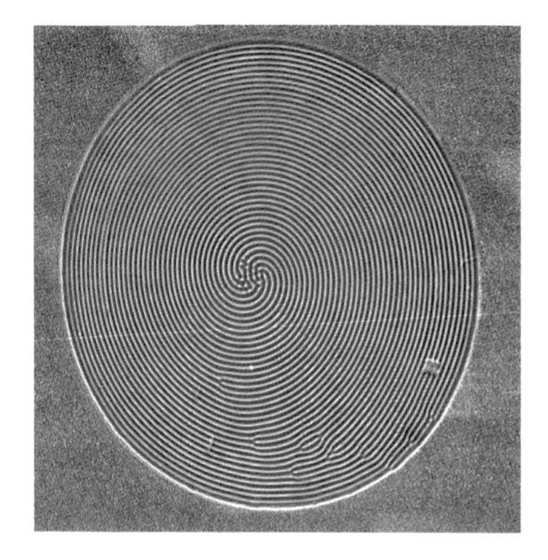

Figure 4 Nine-armed rotating spiral, $\epsilon = 0.19$.

Hexagons and spirals in non-Boussinesq convection

Eberhard Bodenschatz *Cornell University*

Stephen W. Morris *University of Toronto*

John R. de Bruyn
Memorial University of Newfoundland

David S. Cannell and Guenter Ahlers
University of California at Santa Barbara

We present experimentally observed[1–3] flow patterns of non-Boussinesq convection using gaseous, pressurized CO_2 in a cylindrical cell of radius/height ratio 86. The distance from onset of convection is given by the reduced control parameter $\epsilon = \Delta T/\Delta T_c - 1$, where ΔT is the vertical temperature difference and ΔT_c is ΔT at its critical value.

The horizontal planform of the fluid flow is visualized by the shadowgraph technique. White corresponds to cold downflow and dark to warm upflow.

Figure 1 shows the hexagonal flow pattern at onset. The picture is color coded to emphasize the warm upflow (light blue) and the cold downflow (orange) regions.

Figures 2, 3, and 4 show pictures of the full cylindrical cell filled, respectively, with perfectly ordered hexagons, a two armed, and a nine armed clockwise rotating spiral.

This research has been supported by U.S. Department of Energy Grant No. DE-FG03-87ER13738. E. B. was also supported by the Deutsche Forschungsgemeinschaft and J. de B. and S. W. M. by NSERC of Canada.

Keywords

shadowgraph; buoyancy convection; pattern selection.

[1] J. R. de Bruyn, E. Bodenschatz, S. W. Morris, S. P. Trainoff, Y. Hu, D. S. Cannell, and G. Ahlers, "Apparatus for the study of Rayleigh-Bénard convection in gases under pressure", *Rev. Sci. Instrum.* **67**, 2043 (1996)

[2] E. Bodenschatz, W. Pesch, and G. Ahlers, "Recent developments in Rayleigh–Bénard convection," *Ann. Rev. Fluid Mech.* **32**, 709 (2000).

[3] E. Bodenschatz, D. S. Cannell, J. R. de Bruyn, R. Ecke, Y. Hu, K. Lerman, and G. Ahlers, *Physica D* **61**, 77 (1992).

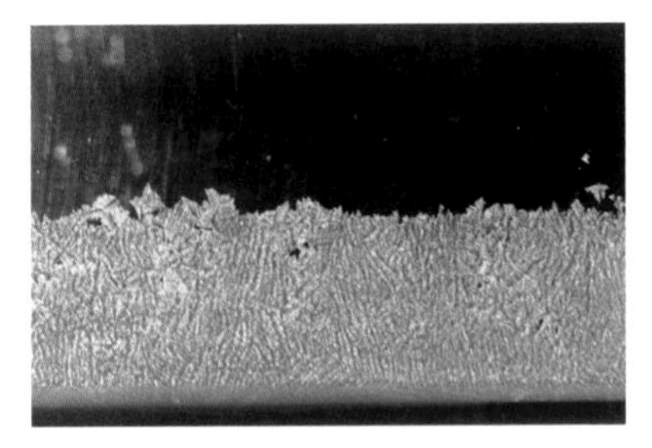

Figure 1(a) 30 min. The dendrites are closely packed owing to the fresh supply of NH_4Cl-rich solution to the freezing front by finger convection. All photos were taken at $f/8$ and 1/15 sec and at 1:1 picture ratio. The actual area depicted in each photo is 24×35 mm.

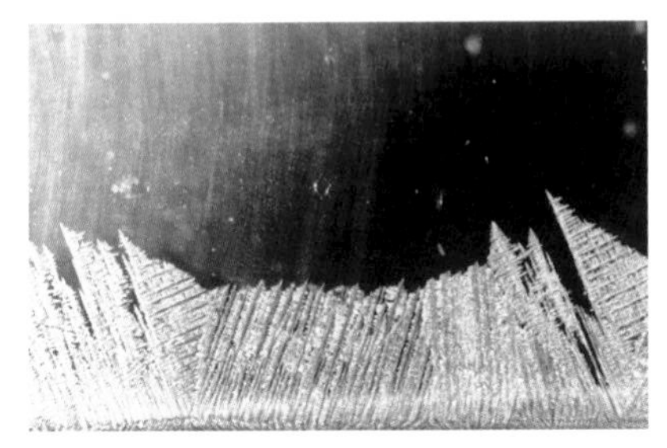

Figure 2(a) 29 min. With higher viscosity, $v/v_0 = 6.49$, and less vigorous convection, the much layer becomes more loosely packed and more structured.

Figure 3(a) 31 min. With the highest viscosity ratio of 27.91, the convection is so weak that the secondary arms of the dendrites have grown to noticeable sizes.

Viscosity effects on the directional solidification of NH_4Cl solution in a Hele-Shaw cell

C. F. Chen *The University of Arizona, Tucson*

A 26% NH_4Cl–H_2O solution was directionally solidified in a Hele-Shaw cell 150×200×1 mm. All sides of the cell were made of Plexiglas with the bottom made of copper which was kept at −21°C. Small amounts of "Natrosol" (hydroxyethylcellulose) were added to the solution to increase its viscosity but without affecting its phase diagram.[1] Experiments were conducted with three samples:

[1] S. Tait and C. Jaupart, *Nature* **338**, 571 (1989).

Figure 1(b) 2 h 30 min. There are four chimneys in the mush layer. The rightmost one is shown here. Note the remelted crystals ascending and descending in the chimney. A large root system has developed near the bottom of the chimney. Cold and lighter fluid rejected by crystallization near the bottom is being fed into the chimney through the root system.

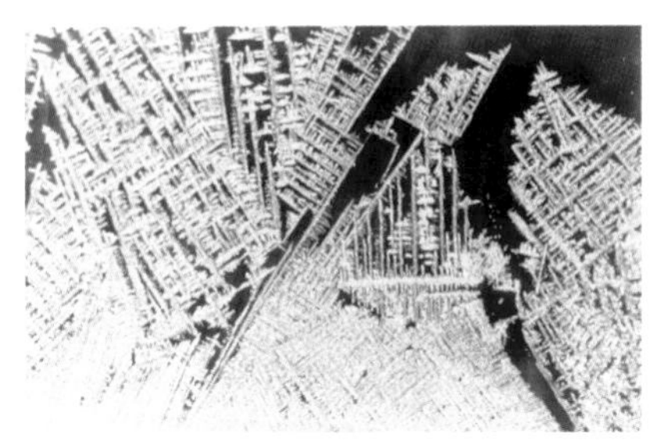

Figure 2(b) 2 h 22 min. Dendritic structure becomes even more loosely packed as the cooling rate is decreased at later times. The large chimney on the right is in the decaying stage. A large chunk of dendritic structure just fell into the channel and is restricting the passage. Note the tiny crystals dropping down from the top.

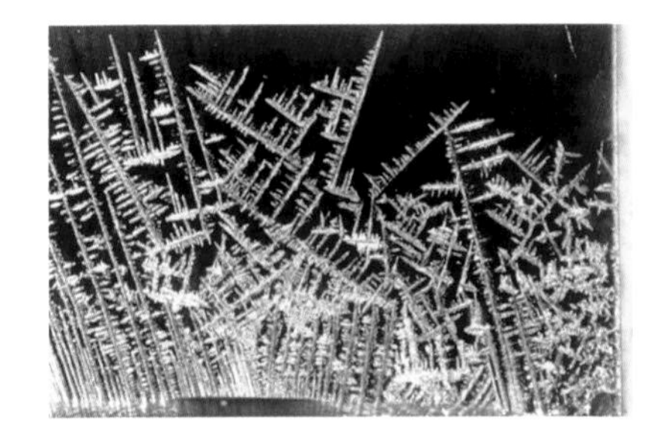

Figure 3(b) 2 h 32 min. No chimneys developed owing to the sparsely packed dendrites. At this time even the tertiary arms of the dendrites are well developed. Dendritic growth is diffusion controlled.

Sample	Natrosol wt%	v/v_0*
1 [Figs. 1(a) and 1(b)]	0.00	1.00
2 [Figs. 2(a) and 2(b)]	0.25	6.49
3 [Figs. 3(a) and 3(6)]	0.50	27.91

(The asterisk indicates $v_0 = 0.0104$ cm²/sec. All measurements were made at 17°C.)

These experiments were performed in the GFD Laboratory of Professor J. S. Turner, Research School of Earth Sciences, Australian National University with the able assistance of T. Beasley, R. Wylde-Browne, and D. Corrigan. The financial support of the NASA Microgravity Science and Applications Division through Grant No. NAG 3-1268 is gratefully acknowledged.

Keywords

finger convection; chimneys; dendrites.

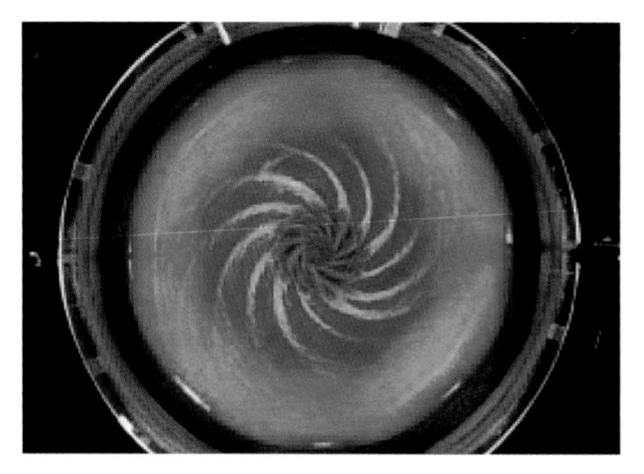

Figure 1

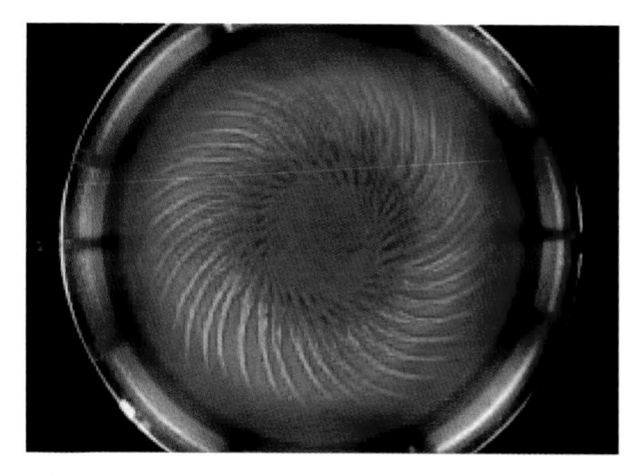

Figure 2

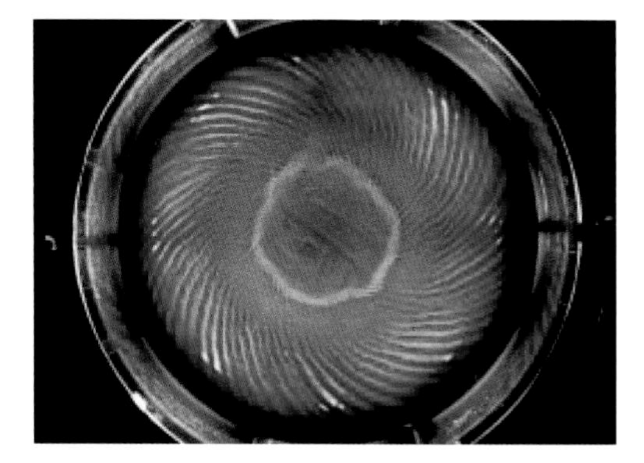

Figure 3

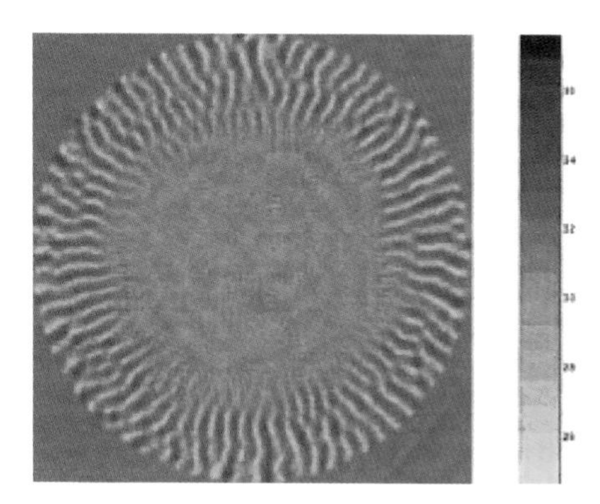

Figure 4

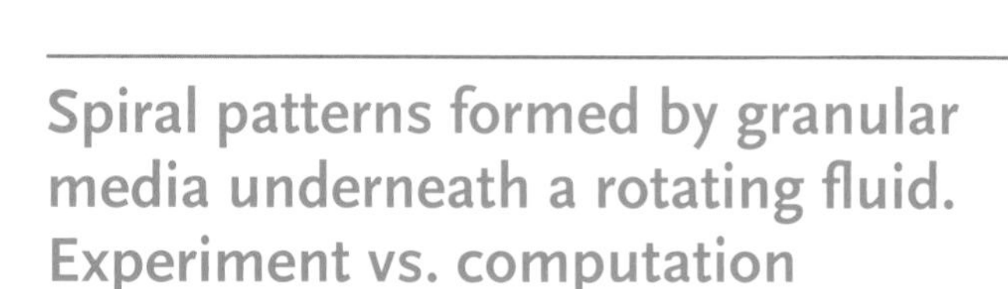

Spiral patterns formed by granular media underneath a rotating fluid. Experiment vs. computation

Farzam Zoueshtiagh and Peter J. Thomas

University of Warwick

Figures 1–3 show spiral patterns formed by small granules on the bottom of a circular, fluid-filled rotating tank. The granule diameter is around 200–250 μm and the diameter of each spiral pattern corresponds approximately to the tank diameter of 1 m. Initially the granules were distributed in a uniform layer across the whole bottom of the tank. The fluid above the layer was in a state of solid-body rotation and, hence, did not move relative to the granules. Pattern formation is initiated when the rotational velocity of the tank is instantaneously increased from its original value ω_0 by a sufficiently large increment $\Delta\omega$ to a higher rate ω_1. The fluid mass inside the tank cannot follow the instantaneous acceleration of the tank. This establishes shear forces between the granule layer and the fluid above it. The granules are set in motion, slide across the bottom of the tank and, within a few seconds, form patterns such as those displayed in Figs. 1–3.

Depending on the experimental conditions we have observed[1] patterns with $7 \leqslant n \leqslant 110$ spiral arms originating from uniform, inner granule patches with associated radii r_0. The data analysis has shown that $n \propto \omega_1^{0.5}/\Delta\omega$ and that $r_0 \propto 1/\Delta\omega$. Recently we have developed a cellular-automaton model[2] which succeeds in generating patterns qualitatively similar to those observed in the experiments. One computational pattern is shown in Fig. 4; the colorbar indicates the height of the granule layer. The model is a suitable generalization of a model previously used[3] to simulate ripple formation in wind-blown sand in a straight channel. Our model, in particular, also reproduces quantitatively the scalings for n and r_0 measured in the experiments. Prompted by the inherent properties of the model, intuitive physical arguments have been advanced[2] which readily yield these scalings as a consequence of the existence of some critical threshold condition. It is argued that this result suggests that the spiral patterns might constitute some type of rotating analogue of sand ripples typically observed on beaches.

[1] P. J. Thomas, "Pattern formation of granules on the bottom of a differentially rotating tank," *J. Fluid Mech.* **274**, 23 (1994).

[2] F. Zoueshtiagh and P. J. Thomas, *Phys. Rev. E* **61** (5), 5588–5592 (2000).

[3] H. Nishimori and N. Ouchi, "Formation of ripple patterns and dunes by wind-blown sand," *Phys. Rev. E* **71**, 197 (1993).

Keywords

simulation; particle ripples.

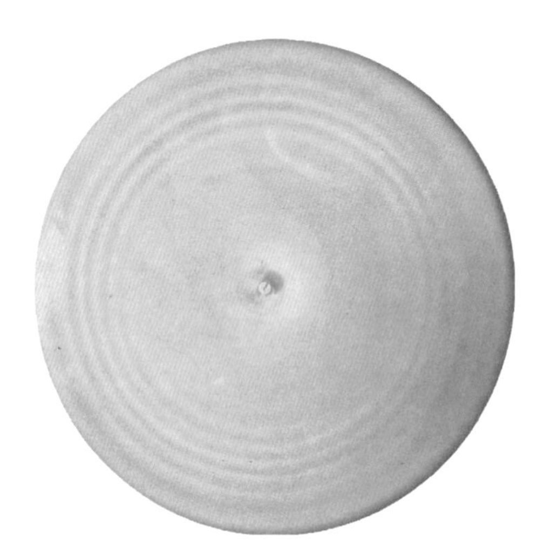

Figure 1

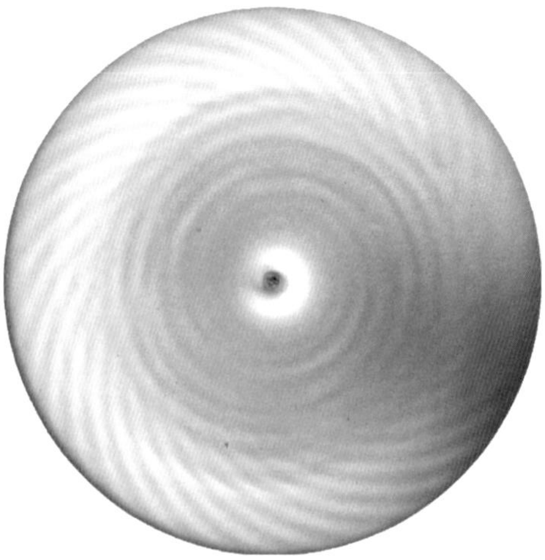

Figure 2

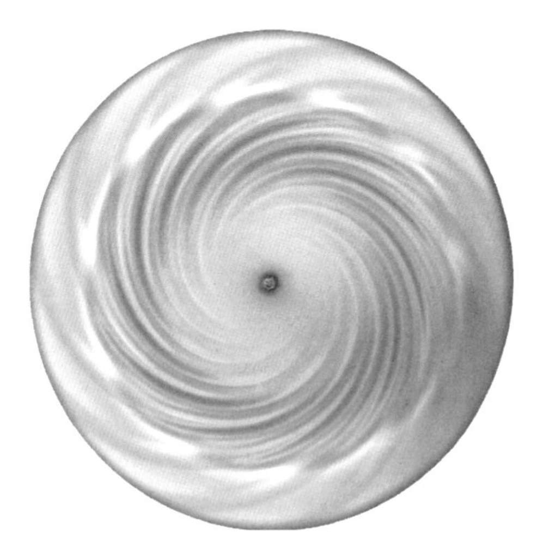

Figure 3

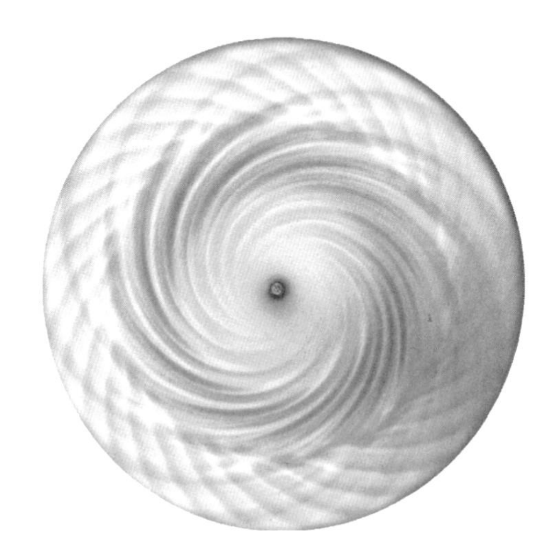

Figure 4

Patterns between two rotating disks

G. Gauthier, P. Gondret, F. Moisy, and M. Rabaud

Laboratoire FAST

These four pictures show the instabilities of the flow between two rotating disks enclosed by a cylinder. The disks, 28 cm in diameter, are 7 mm apart. The fluid is a mixture of water and glycerol, seeded with anisotropic flakes to allow visualization. The Reynolds number (built with the gap thickness) ranges typically from 10 to 200 for these pictures. When the disks rotate in the same direction, or in low counter-rotation, we first observe circles[1] (Fig. 1) propagating towards the center. Increasing the disk velocities, we see a spiral pattern (Fig. 2), rotating in the direction of the disks (here clockwise). These two patterns are due to boundary layer instabilities. When the disks rotate in opposite direction, another spiral pattern slowly appears (Fig. 3), which fills the whole gap.[2] This new pattern arises through a super-critical bifurcation. Increasing the Reynolds number, the two spiral patterns can coexist (Fig. 4).

[1] G. Gauthier, P. Gondret, and M. Rabaud, "Axisymmetric propagating vortices in the flow between a stationary and a rotating disk enclosed by a cylinder," *J. Fluid Mech.* **386**, 105 (1999).

[2] G. Gauthier, P. Gondret, F. Moisy, and M. Rabaud, "Instabilities in the flow between co- and counter-rotating disks," *J. Fluid Mech.* **473**, 1–21 (2002).

Keywords
flake reflection; boundary-layer instability; shear layer.

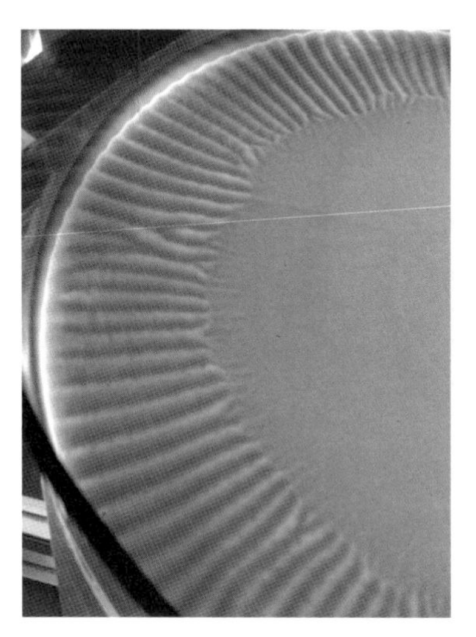

Figure 1(a)

Figure 1(b)

Figure 2(a)

Figure 2(b)

Figure 2(c)

Sand ripples in a rotating tank

Marc Fermigier and Patrice Jenffer
ESPCI

The generation of ripples on sand dunes and sediment beds by a fluid flow has received a renewed attention recently.[1,2] Here we study the pattern of sand ripples formed in a circular water tank (90 cm in diameter) in oscillatory (Fig. 1) and continuous (Fig. 2) motion. With an oscillatory shear, the ripples are essentially radial with some defects in the pattern to adjust the wavelength. The ripples start at the periphery of the tank where the shear stress is larger [Fig. 1(b)]. At the beginning of the experiment, there is a coexistence between small amplitude "rolling grain ripples" and large amplitude "vortex ripples" [Fig. 1(a)].

We also perform continuous shear experiments where the layer of water is confined between the bed of sand, rotating with the tank, and an upper, fixed, Plexiglas plate [Fig. 2(a)]. Small depressions in the otherwise flat bed of sand are used to initiate a system of ripples [Fig. 2(a)]. From the deepest initial depression, a pattern of vortex ripples developed downstream, with a wavelength decreasing in the downstream direction and a segmentation of the ripple pattern in the spanwise direction [Fig. 2(b)]. A detail of the downstream end of the pattern is shown on Fig. 2(c).

Keywords
sand ripples; oscillatory shear.

[1] A. Stegner and J. E. Wesfreid, *Phys. Rev. E* **60**, R3487 (1999).

[2] M. A. Scherer, F. Melo, and M. Marder, *Phys. Fluids* **11**, 58 (1999).

Figure 1

Figure 2

Figure 3

Figure 4

The thistle crown

E. Tan and S. T. Thoroddsen

University of Illinois at Urbana-Champaign

This flower-like pattern evolves after the impact of a dyed water drop onto a thin layer of glycerin. The impact breaks the drop into an outer ring and a center bead. Subsequently, Marangoni stresses arise at the free surface driving strong vortices inside the water layer. These stresses arise from the surface tension variations due to the miscibility of the two liquids. Azimuthal undulations of the interface are reinforced by enhanced surface-tension gradients, leading to the instability demonstrated by the leafing. The drop diameter is about 5 mm and it is released from 33 cm height. The glycerin layer is about 2 mm deep. Figure 1 shows the drop fluid during the impact, demonstrating the formation of an internal crown, along which the drop rips into the inner and outer regions. The patterns shown in Figs. 2–4 are separated by approximately 5 s. The total diameter of the pattern in Fig. 4 is about 1.5 cm.

Keywords

Marangoni stresses; surface tension gradients; fluorescent dye.

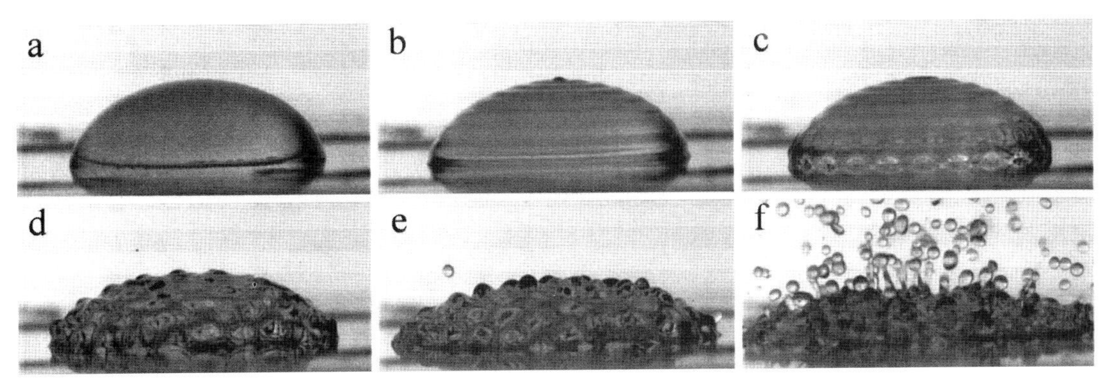

Figure 1 A water drop (0.1 ml) forced by a slowly ramped vibration at 903 Hz: (a) unforced, (b) axisymmetric waves, (c) coupling of the axisymmetric and azimuthal waves, (d) complex pre-ejection state, (e) ejection onset, and (f) atomization.

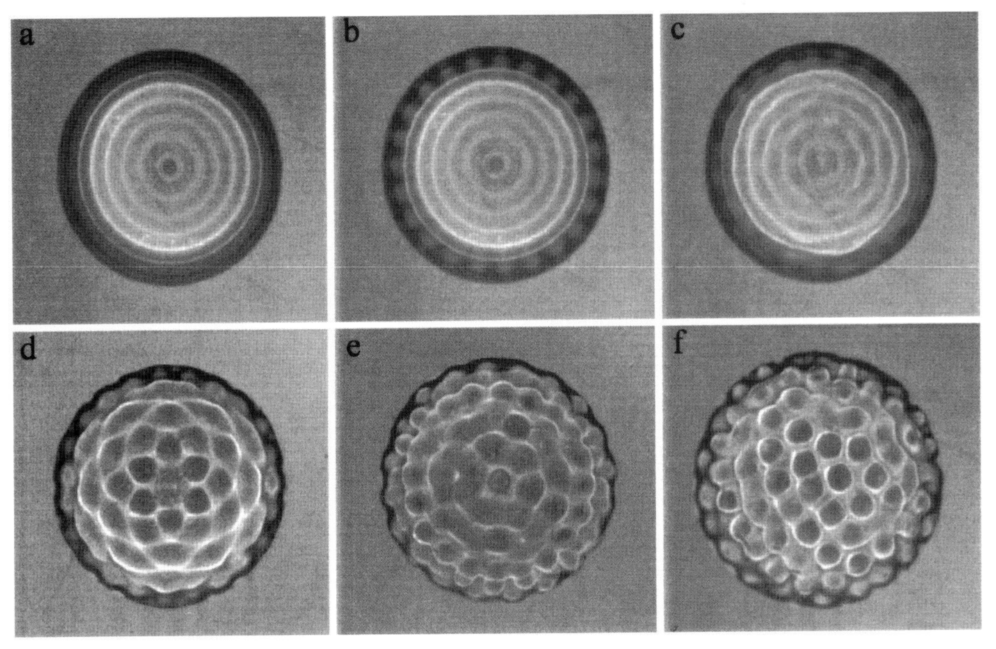

Figure 2 The transition modes of a water drop (0.1 ml) forced at 903 Hz: (a) axisymmetric wave, (b) azimuthal waves, (c) azimuthal breakdown, (d) lattice pattern, (e) lattice breakdown, and (f) the pre-ejection mode.

Mode shapes of a sessile drop in forced vibration

Bojan Vukasinovic, Ari Glezer, and Marc K. Smith

Georgia Institute of Technology, Atlanta, GA

Earlier work has shown that a sessile drop forced by a vibrating diaphragm undergoes several transitions prior to ultimate atomization.[1] The present work focuses on the evolution of waves on the drop surface (Fig. 1). At low forcing amplitudes, axisymmetric waves are induced by the motion of the diaphragm (Fig. 1b). As the forcing amplitude is increased, an azimuthal instability appears along the contact line of the drop and couples with the existing axisymmetric waves (Fig. 1c). A further increase in the forcing amplitude results in complex free-surface waves (Fig. 1d) and ultimately leads to the ejection of secondary droplets (Fig. 1e) and atomization (Fig. 1f). The free surface (and internal) motions associated with these transitions result in radial spreading of the drop.

A closer look at the drop free-surface (Fig. 2) reveals that the forced drop undergoes three primary transitions namely, from axisymmetric to azimuthal waves (Fig. 2b), then to a newly observed lattice-like pattern (Fig. 2d) having a 6-fold symmetry, and then to a chaotic state that is the precursor to ejection and atomization (Fig. 2f).

Keywords

Faraday forcing; droplet ejection; atomization.

[1]B. Vukasinovic, A. Glezer, and M. K. Smith, "Vibration-induced droplet atomization," *Phys. Fluids* **12**, S12 (2000).

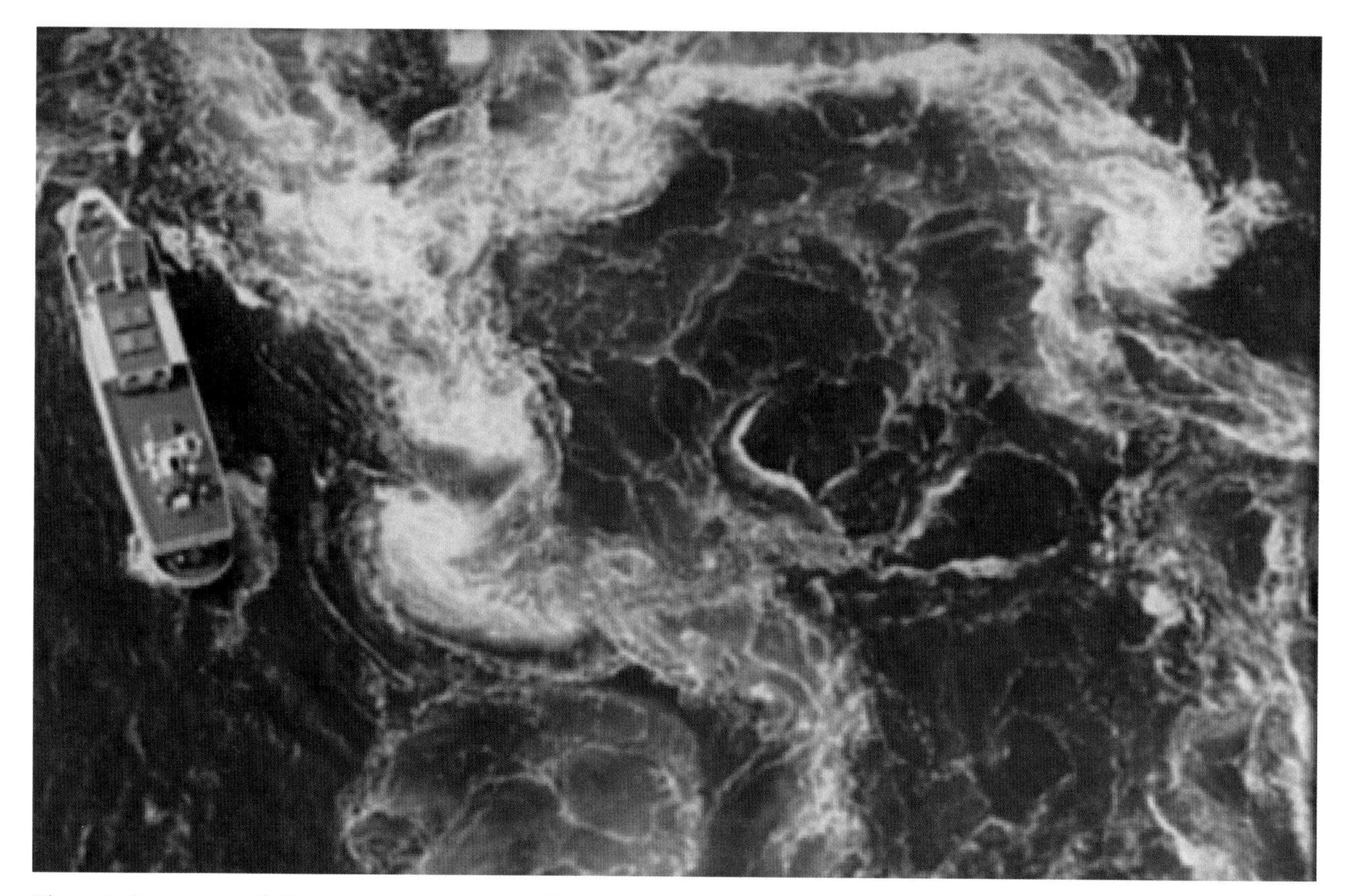

Figure 1 Photograph of *Naruto* taken from a Yomiuri Shimbun helicopter during early March, 1996, by Masafumi Nanjo of the *Daily Yomiuri*, Tokyo, Japan.

Naruto: past and present*

Norman J. Zabusky and Wesley Townsend
Laboratory for Visiometrics and Modeling
Rutgers University

Keywords
maelstrom; whirlpool; tidal vortex.

The Naruto strait contains a tidal current whose edges constitute a gigantic cascading "maelstrom." It is not far from Kobe, Japan and lies between the city of Naruto on Shikoku Island and the island of Awaji (Hyogo Prefecture) and connects the higher Seto Inland Sea (Setonaikai) and the lower Osaka Bay.[1] The rushing tidal current is a sloping surface jet (exceeding 5 m/s) in-or-out between the two seas. The phenomenon is strongest during the first hour of every 6 h and 25 min period, particularly at full moons in the early spring, when Nanjo's photograph was taken (early March, 1996). A bridge now spans Naruto to Awaji and was finished in March of 1985.

In the photograph we see a near-vertical aerial view of a sight-seeing boat close to the southwestern (counter-clockwise-vortex) edge of the cascade. Dominant vortex structures may be 30 m in diameter. We await the quantification of the stratified turbulence of this natural wave-vortex system.

* This page summarizes the Naruto images collected by N. J. Zabusky during his visit to Japan in 1996. The photograph above was provided by Masafumi Nanjo of the Daily Yomiuri, Tokyo, Japan and is printed with his permission. The photograph was submitted as the poster, "Naruto '96" to the 1996 Gallery of Fluid Motion at the American Physical Society Division of Fluid Dynamics annual meeting in November, 1996. Another submission was a video "NARUTO: Past and Present," created by Norman J. Zabusky and Wesley Townsend. It was based on the 19th-century Ukiyo-e print by Ando Hiroshige and an NHK Tokushima video made during the bridges inauguration ceremony in March of 1985 and Nanjo's print. The video was provided by Professor K. Ishii of the Department of Applied Physics at Nagoya University. N. J. Zabusky acknowledges the gift of these images. Many of them can be seen on our URL home page, http://caip.rutgers.edu/vizlab.naruto.html

[1] For sightseeing information, see Jay Gluck, Sumi Gluck, and Garet Gluck, *Japan Inside Out* (Personally Oriented, Ltd., Ashiya, Japan, 1992).

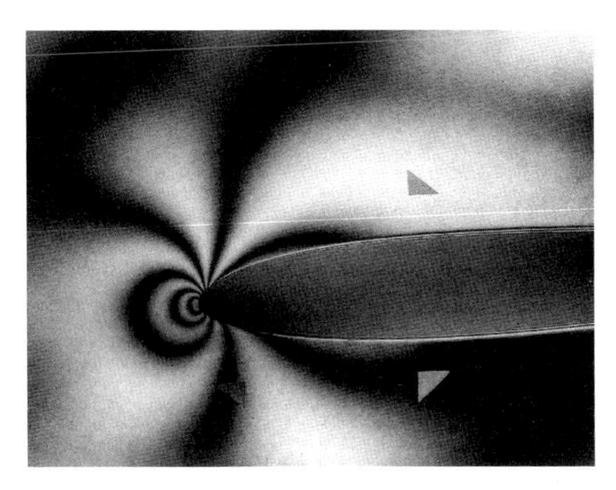

Figure 1(a) a ≅ 0°.

Figure 1(b) a ≅ 7°.

Figure 1(c) a ≅ 14°.

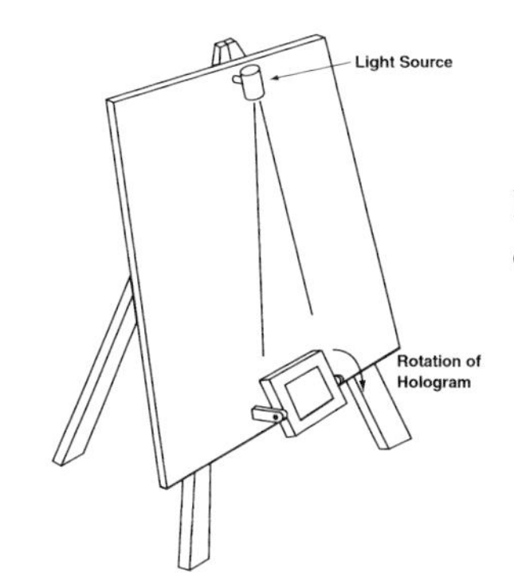

Figure 2 Hologram poster display.

Dynamic stall – a holographic animation of compressible flow interferograms

R. A. Hess, N. J. Brock, and B. J. Weber

Aerometrics, Inc.

L. W. Carr

NASA Ames Research Center

M. S. Chandrasekhara

US Navel Postgraduate School

Rapid pitching of an airfoil past the angle at which it would stall in steady flow creates a dynamic delay of stall known as "dynamic stall," a phenomenon which is under study[1] because it presently limits the forward speed and maneuverability of helicopters and influences the potential maneuverability of fighter aircraft. Analysis of dynamic stall in compressible flows has been limited by the difficulty of obtaining accurate visualization of this complex flow field. Quantitative instantaneous visualization of this flow field has now been obtained through use of a point diffraction interferometry technique which was constructed by modifying an existing schlieren system.[2] Interferograms of the flow over an oscillating airfoil at $M = 0.30$, $\alpha = 10° + 10° \sin(kt)$, $k = \omega c/2U = 0.05$ can be seen in Figs. 1(a)–1(c). The dark lines or "fringes" in each interferogram represent contour lines of constant density, thus offering a complete quantification of this unsteady flow field at each instant. Dynamic stall is indicated by deflection of the fringe pattern away from the airfoil upper surface, as can be seen in Fig. 1(c).

This quantitative flow visualization is a particularly powerful method for conveying information and providing understanding of the critical flow field phenomena. However, for dynamic flows, sequences of moving images are often much more effective than the static images that are often presented. To display the dynamic stall phenomenon, holographic techniques were used to combine the advantages of moving images with the convenience of poster format presentation.

In this holographic poster display, 15 flow visualization images of the developing flow around an airfoil during dynamic stall are displayed on a single holographic plate. The hologram is attached to a poster with a pivoting mount (Fig. 2). A small incandescent bulb is mounted to the top of poster to ensure proper illumination. The holographic technique places all of the 15 individual images on a single holographic plate, but presents only one image at a time. By tilting the holographic plate a few degrees, the previous image disappears and is replaced at the same location by the next image in the sequence. A moving, animated sequence of images is produced by steadily rotating the plate through its entire range of motion. The sequence can be viewed one frame at a time, at any desired speed, or in reverse order, by altering the rotation rate and direction of the plate motion. This new presentation technique thus permits the viewer to study the developing flow in detail.

Keywords

interferometry; hologram; airfoil.

[1] L. W. Carr, M. S. Chandrasekhara, S. Ahmed, and N. Brock, "A Study of Dynamic Stall Using Real-Time Interferometry", *Journal of Aircraft* **31** (4), 991–993 (1994).

[2] N. J. Brock, M. S. Chandrasekhara, and L. W. Carr, "A real-time interferometry system for unsteady flow measurements," ICIASF Conference, Rockville, Maryland, 25–28 October 1991.

4 Drops and bubbles

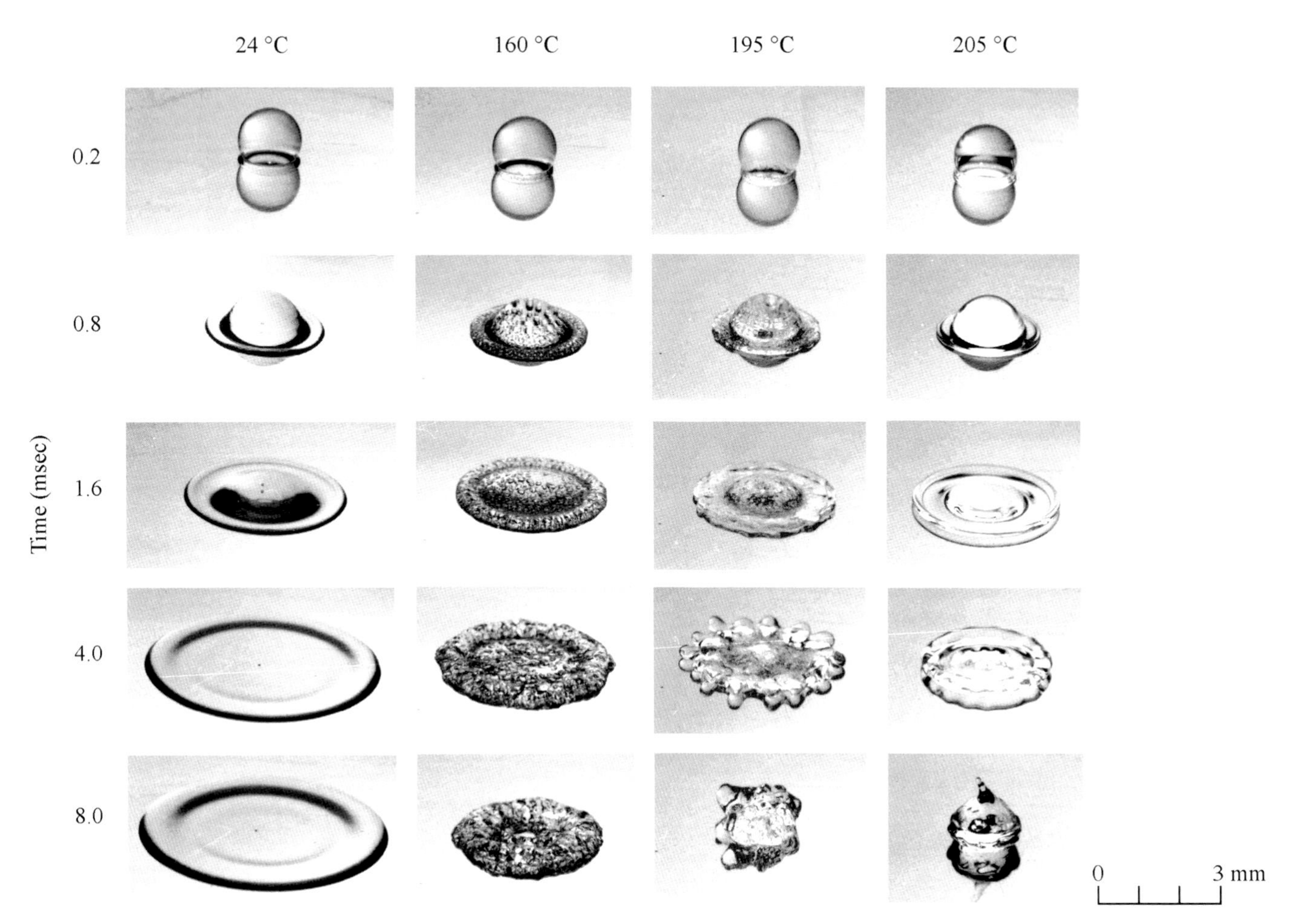

Figure 1

The collision of a droplet with a solid surface

S. Chandra and C. T. Avedisian

Cornell University

The photographs displayed above show the impact, spreading, and boiling history of *n*-heptane droplets on a stainless steel surface. The impact velocity, Weber number, and initial droplet diameter are constant (values of 1 m/s, 43 and 1.5 mm respectively), and the view is looking down on the surface at an angle of about 30°. The photographs were taken using a spark flash method[1] and the flash duration was 0.5 μs. The dynamic behavior illustrated in the photographs is a consequence of varying the initial surface temperature.

[1] S. Chandra and C. T. Avedisian, "On the collision of a droplet with a solid surface," *Proceedings of the Royal Society of London* **A 432**, 13–41 (1991).

The effect of surface temperature on droplet shape may be seen by reading across any row; the evolution of droplet shape at various temperatures may be seen by reading down any column. An entrapped air bubble can be seen in the drop when the surface temperature is 24°C. At higher temperatures vigorous bubbling, rather like that of a droplet sizzling on a frying pan, is seen (the boiling point of *n*-heptane is 98°C) but the bubbles disappear as the Leidenfrost temperature of *n*-heptane (about 200°C) is exceeded because the droplet become levitated above a cushion of its own vapor and does not make direct contact with the surface. The droplet shape is unaffected by surface temperature in the early stage of the impact process ($t \leq 0.8$ ms) but is affected by temperature at later time (cf. $t \geq 1.6$ ms) because of the progressive influence of intermittent solid–liquid contact as temperature is increased.

Keywords

cavitation bubble; Leidenfrost effect; surface tension.

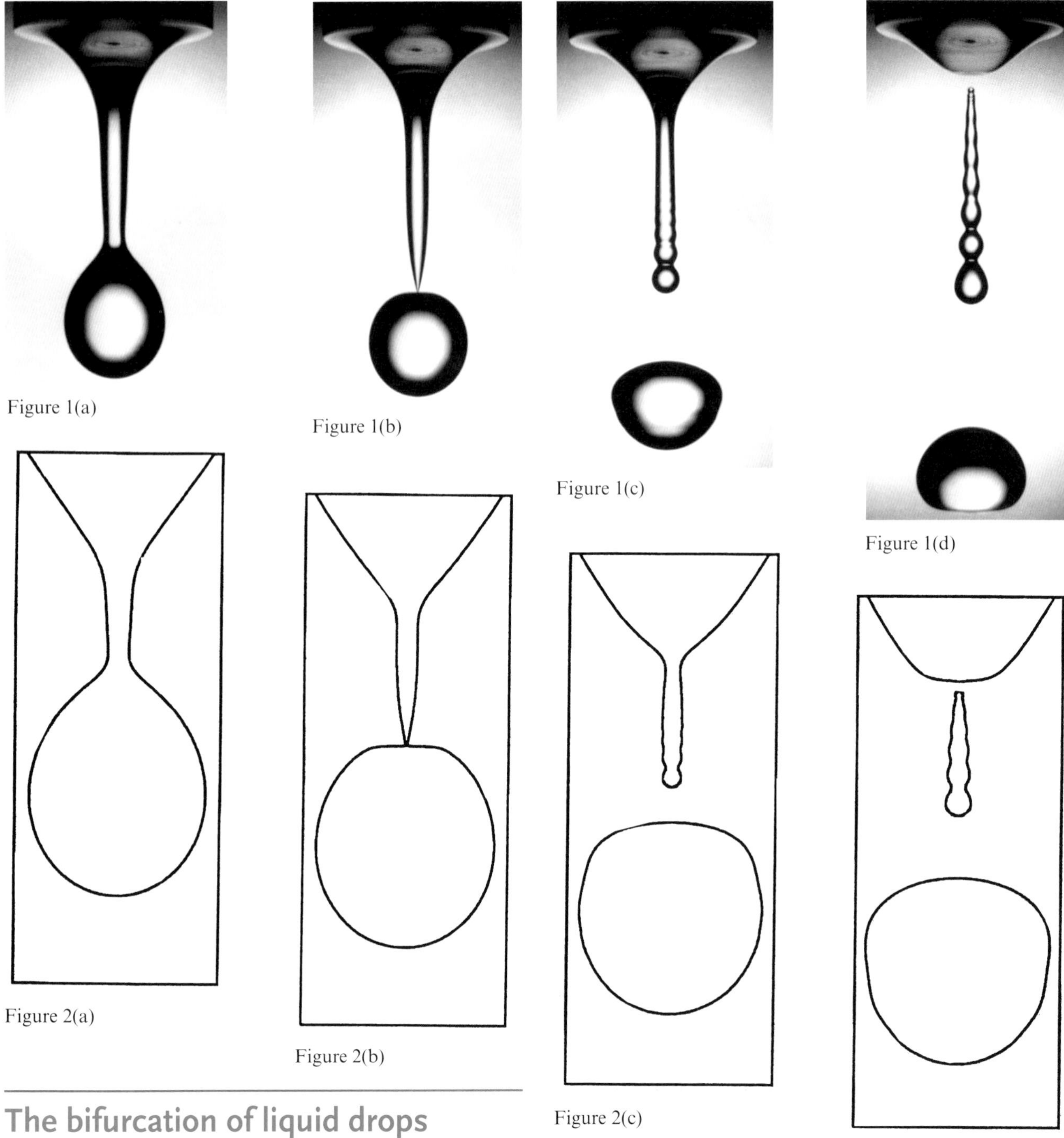
Figure 1(a) Figure 1(b) Figure 1(c) Figure 1(d)

Figure 2(a) Figure 2(b) Figure 2(c) Figure 2(d)

The bifurcation of liquid drops

Michael P. Brenner, X. D. Shi, Jens Eggers, and Sidney R. Nagel

Massachusetts Institute of Technology

Keywords

surface tension; dripping faucet; droplet pinchoff; simulation and experiment.

The breakup of fluid drops is one of the simplest and most ordinary examples of a hydrodynamic singularity, in which physical quantities diverge in a finite amount of time. These pictures document our experiments and computer simulations of a water drop falling from a nozzle.[1,2]

In the top panel we show a sequence of photographs in which the drop first breaks at the bottom and then near the nozzle. The second row shows a numerical simulation of the process using modified long-wavelength equations.[3]

[1]X. D. Shi, M. P. Brenner, and S. R. Nagel, "Cascade of structure in a drop falling from a faucet," *Science* **265**, 219 (1994).

[2]M. P. Brenner, J. Eggers, K. Joseph, S. R. Nagel, and X. D. Shi, "Breakdown of scaling in droplet fission at high Reynolds number," *Phys. Fluids* **9**, 1573 (1997).

[3]J. Eggers and T. F. Dupont, "Drop formation in a one-dimensional approximation of the Navier–Stokes equation," *J. Fluid Mech.* **262**, 205 (1994).

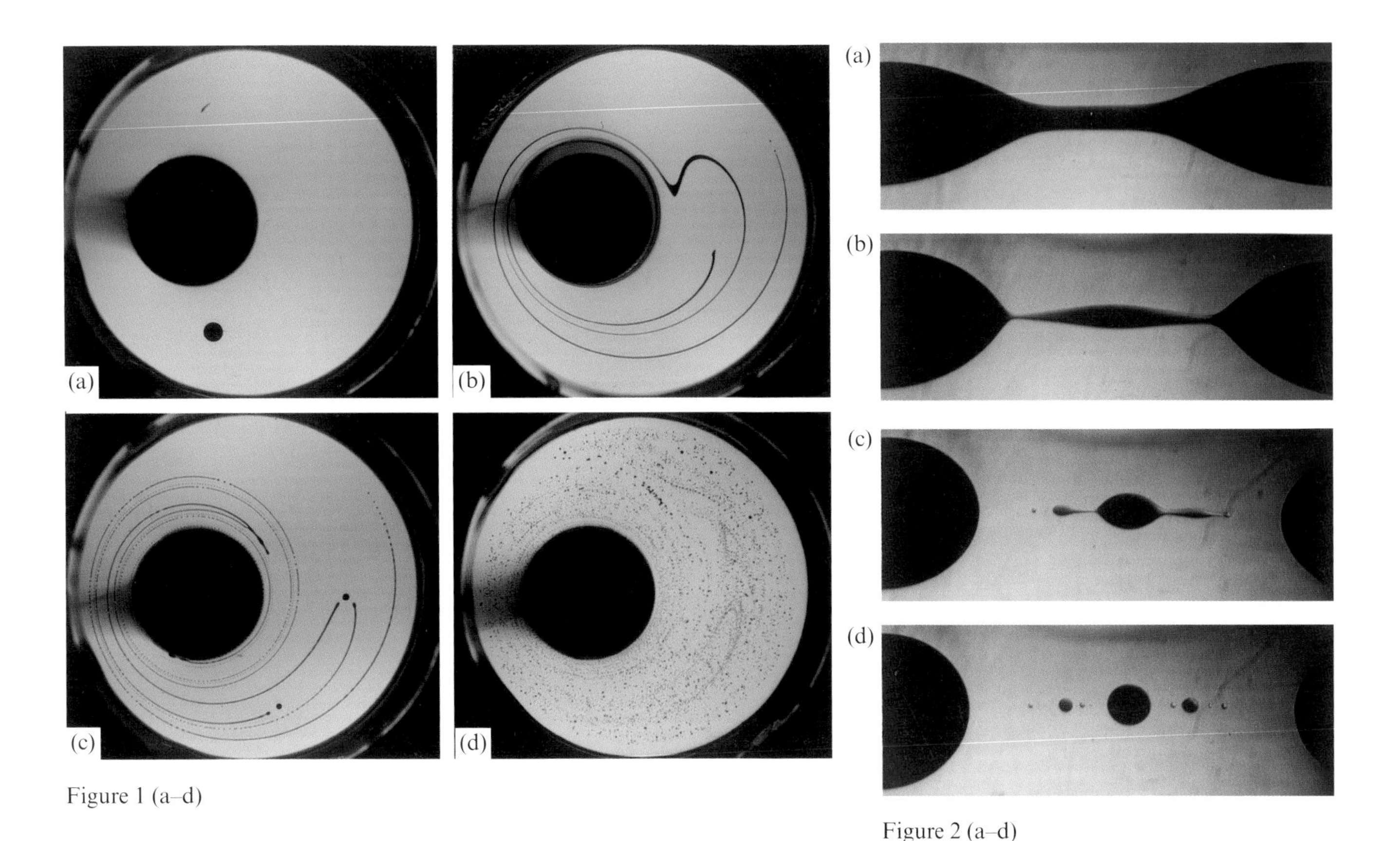

Figure 1 (a–d)

Figure 2 (a–d)

Breakup and dispersion of highly stretched droplets in a chaotic flow field

M. Tjahjadi and J. M. Ottino

University of Massachusetts at Amherst

These photographs show the time evolution – stretching and breakup – of immiscible drops in a two-dimensional time-periodic flow.[1] The initial condition in Fig. 1(a) is a drop of pale oil immersed in corn syrup ($\mu = 38\ 000$ cps) filling the gap between the two eccentric cylinders. The viscosity ratio between the drop fluid and the bulk fluid is about 0.38 and the interfacial tension is 9.5 dyn/cm. The typical Reynolds number for a drop of radius 0.1 cm is $O(10^{-4})$; the Reynolds and Strouhal numbers of the macroscopic flow are $O(10^{-2})$ and $O(10^{-1})$. The flow is induced by moving the inner ($r_{in} = 1$ in.) and outer cylinders ($r_{out} = 3$ in.) periodically, in clockwise and counterclockwise directions, respectively. The total displacement of each cylinder during a period is kept the same, i.e., $\theta_{in} r_{in} = \theta_{out} r_{out}$, and the wall velocity of each cylinder is 1.0 cm/sec. Figures 1(a)–1(d) correspond to $\theta_{out} = 2\pi$. At the end of one period the flow is stopped; the first drops form at the end of the filament and at the folds; concurrently, in $O(10$ sec), capillary instabilities grow and lead to total fragmentation [Fig. 1(c)]. The process is repeated and the remaining "larger" drops undergo further breakup [Fig. 1(d)]; note that the large fold in Fig. 1(c) is also recognizable in Fig. 1(d). Figures 2(a)–2(d) show a magnification of a small segment of filament undergoing breakup; the asymmetry is probably due to small variations in the radius of the thread.

This work was supported by the U.S. Department of Energy, Office of Basic Energy Sciences.

Keywords
chaotic mixing; droplet breakup; interfacial tension.

[1] P. D. Swanson and J. M. Ottino, *J. Fluid Mech.* **213**, 227 (1990).

Figure 1

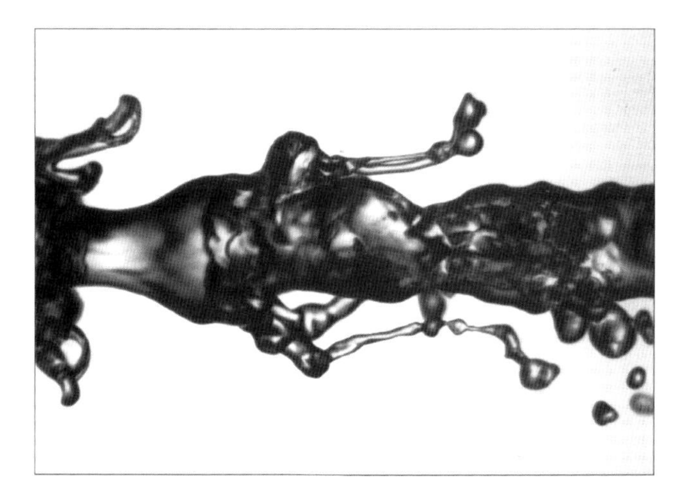

Figure 2

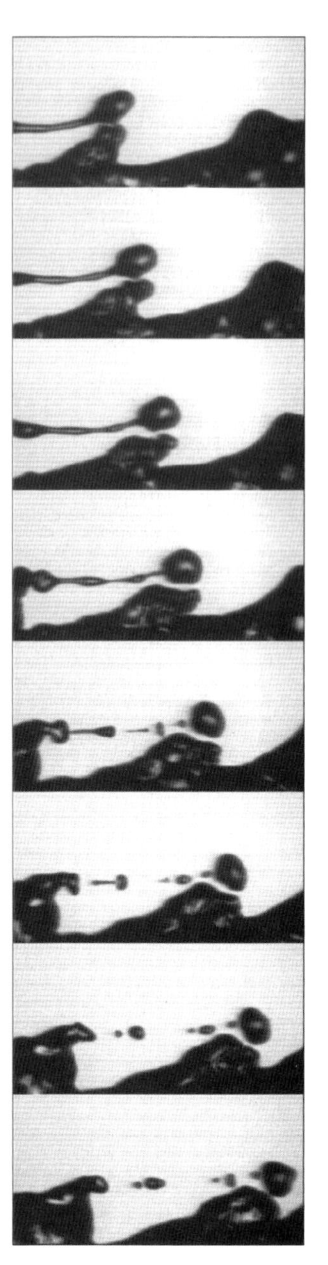

Figure 3

Ligament mediated drop formation

Ph. Marmottant and E. Villermaux

LEGI, Grenoble and IRPHE, Marseille

The atomization of a liquid jet by a fast coaxial gaseous stream involves the formation of ligaments which elongate in the rapid gas stream and eventually break into disjointed droplets.

A longitudinal shear instability first develops (Fig. 1), controlled by the initial gas shear layer thickness.[1] The passage of these waves confers to the liquid transient accelerations perpendicular to the interface which trigger a transverse instability of a Rayleigh–Taylor type from which ligaments form (Fig. 2).

After the breakup of each ligament (Fig. 3), the distribution of the drop sizes in the spray is a decaying exponential. The broad distribution of the ligament volumes and of the droplet sizes issued from each ligament is responsible for the broad statistics of the final spray distribution in this one-step, noncascading atomization process.

[1] E. Villermaux, "Mixing and spray formation in coaxial jets," *J. Prop. Power* **14**, 807 (1998).

Keywords

jet atomization; spray generation.

Figure 1 Rapid atomization of a 0.1 ml water droplet forced by stepped actuation at 1080 Hz. The frame rate is 2000 fps.

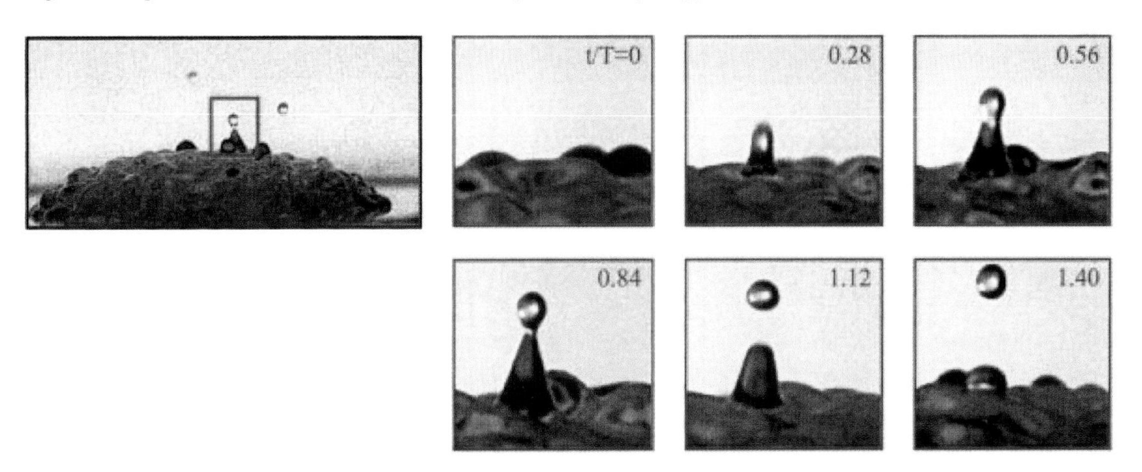

Figure 2 A single ejection from the free surface of a 0.1 ml water droplet forced by slowly ramped actuation at 840 Hz. The sequence of zoomed-in images is recorded at 3000 fps.

Vibration-induced droplet atomization

Bojan Vukasinovic, Ari Glezer, and Marc K. Smith

Georgia Institute of Technology, Atlanta, GA

The atomization of a liquid drop placed on a round vibrating diaphragm is investigated experimentally. The diaphragm is driven in its fundamental axisymmetric mode at or near resonance by a smaller piezoceramic actuator disk that is bonded at the center of the diaphragm and operates in its radial (shearing) mode. Figure 1 is a sequence of high-speed video images that show a 10 mm diameter drop being atomized within 400 ms. The drop surface undergoes a hierarchy of instabilities that lead to the formation of surface waves and the atomization or bursting is a result of the rapid ejection of small secondary droplets from the crests of the these waves. The atomization is triggered along the circumference of the primary drop near the contact line by a strong azimuthal instability. The evolution and rate of ejection of these secondary droplets depend on the coupled dynamics of the primary drop and the vibrating diaphragm.

A single secondary-droplet-ejection event is shown in Figure 2. These events are apparently initiated by the interplay of the spatial modes on the highly contorted surface of the primary drop. The vertical motion of the liquid spike is preceded and directly initiated by the collapse of a free-surface crater or depression. The secondary droplet separates from the liquid spike by a capillary pinchoff mechanism. If it has enough momentum it will then be ejected from the primary drop.

Keywords

droplet ejection; Faraday forcing; capillary pinchoff.

Figure 1

Atomization of a planar liquid sheet

A. Lozano
LITEC/CSIC, Spain

C. J. Call and C. Dopazo
Universidad de Zaragoza, Spain

This photograph presents the breakup of a liquid sheet formed in a planar atomizer. The liquid (water) sheet thickness was 0.9 mm, and the length was 80 mm. Air was coflowing along both sides of the water sheet through 1 cm apertures. To obtain instantaneous images, a 0.5 μs flash lamp was used to freeze the water motion allowing visualization of the sheet and its disintegration. Experiments were performed for water mass flow rates ranging from 20 g/s (~0.28 m/s) to 180 g/s (2.5 m/s), and air flow rates from 20 g/s (12.5 m/s) to 120 g/s (75 m/s).[1] The photograph shows a side view of the breakup process obtained with back illumination for water velocity of 1 m/s and air velocity of 30 m/s. The sheet flapping caused by the Kelvin–Helmholtz instability is clearly visible, although the ligament formation caused by the streamwise vorticity cannot be discerned in this view. The image is presented upside down because of its curious resemblance to a night landscape.

[1] A. Lozano, C. J. Call, and C. Dopazo, "An experimental and numerical study of the atomization of a planar liquid sheet," 6th International Conference on Liquid Atomization and Spray Systems, Rouen, France, July 1994.

Keywords

liquid sheet disintegration; Kelvin–Helmholtz instability.

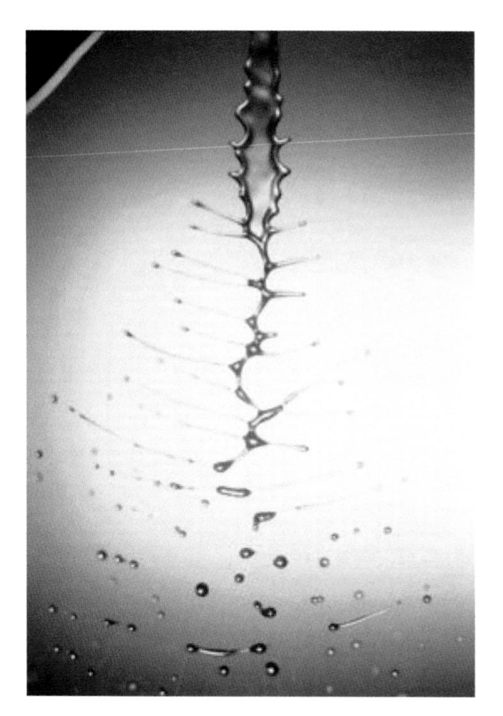

Figure 1

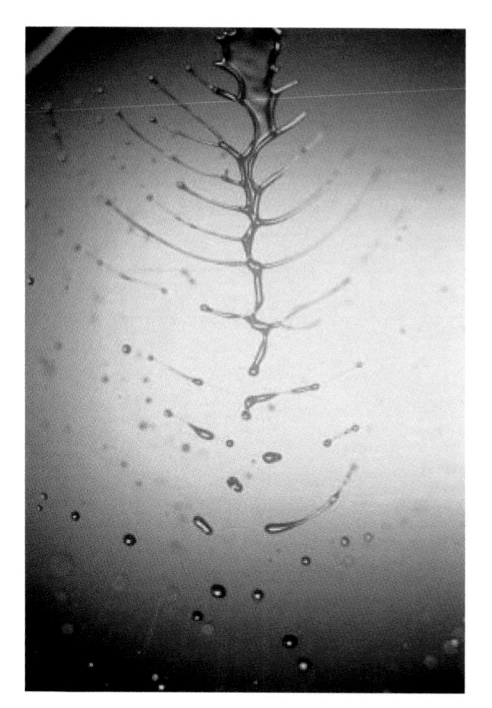

Figure 2

Formation of a net-negative charged liquid spray

Dimitris E. Nikitopoulos *Louisiana State University*

Arnold J. Kelly *CIC, Inc.*

The figures show two instantaneous realizations of electrostatic disruption of room-temperature parafinic white oil. The oil is issuing at 5 mL/s through a smooth 1 mm diameter circular orifice (not shown) after having been negatively charge injected to a mean charge density of 0.15 Cb/m^3.

The liquid is charged by a submerged electrode, which is positioned immediately upstream of the grounded orifice through which the fluid issues. In absence of charge injection the liquid would exit as a glassy smooth cylindrical stream. The elegant filamentary structure and subsequent droplet development is purely electrostatic. No mechanical or aerodynamic forces are involved.

Droplet development starts with collapse of the cylindrical charged column into the thick-rimmed, ribbon-like structure seen at the top of the images. The charged cylindrical column of fluid is unstable to small perturbations of the circular cross section of the column. Upon small distortion, the charge, and therefore the fluid that carries it, tends to rapidly migrate to the extremities pulling the column into a thin sheet. The concentration of charged fluid at the extremities accounts for the thick rim of the ribbon. The rim, in turn, is unstable to longitudinal waves. Again the charge rushes to the extremities, this time the crests of the waves, stretching them into rapidly growing, periodically occurring thin filaments. As the charged fluid moves toward their tips, the filaments are typically capped by growing fluid lumps. These subsequently break off in the form of mutually repelling and rapidly dispersing droplets. As seen toward the bottom of the images, occasionally the filaments themselves break off from the maternal ribbon and subsequently break up into droplets on either end, by means of the same basic mechanism.

Description of the large-scale electrohydrodynamic motion has yet to be attempted. However, the resulting droplet spray is quantitatively described by statistical thermodynamics. The system is well mixed and in its most probable, maximum entropy state. The description reveals the existence of a first order phase transition in charging level at about a micron. Mean charge on droplets larger than a micron is simply the ratio of mean droplet size to the first Bohr radius. One interpretation of this independence of droplet size on fluid properties is that spray behavior is determined by the physics of the two-dimensional surface charge; the transition being the point at which the surface charge freezes into an electron crystal. The droplets in the image are most probably coated with electron ice crystals.

The electrostatically atomized working fluid is EXXON Marcol 87 with viscosity ~35 cp, surface tension =0.031 N/m, density=850 kg/m^3, and electrical conductivity ~0.1 cu.

Keywords

electrostatic spray; spray generation; droplet formation.

Figure 1

Figure 2

Figure 3

Figure 4

Explosive boiling of a droplet at the superheat limit

D. Frost and B. Sturtevant
California Institute of Technology

The photographs demonstrate the explosive boiling of droplets of ether immersed in glycerol. The temperature is 150 °C and the ambient pressure is 3 bar. At these conditions, homogeneous nucleation occurs spontaneously at the limit of superheat and a smooth vapor bubble grows relatively slowly and stably within each drop. The ether droplet shown in Fig. 1 was photographed after it had been evaporating stably for over 7 msec. Only a small amount of liquid remains in the shape of a thin cap that is visible in the upper left portion of the photograph. The thick black band surrounding the vapor region is caused by optical refraction. The bubble has an average diameter of 4 mm.

Incipient instability waves (with a length scale of 100 μm) have developed on the evaporating surface of the drop in Fig. 1 and give it a distinctive "orange-peel" appearance. Figure 2 shows a partially vaporized ether droplet photographed 115 μsec *after* naturally occurring temperature fluctuations had initiated violent boiling near the edge of the remnant liquid region. The clamshell-shaped unstable region rapidly spreads out radially, consuming the remaining liquid.

Figure 3 contains a profile view of an ether bubble (taken 120 μsec after the onset of unstable boiling) showing a miniature eruption within the bubble. During unstable boiling, very fine liquid droplets are torn from the evaporating interface and entrained in the flow leaving the interface. The average velocity of this transient two-phase jet is about 50 m/sec, and the corresponding mass flux is several orders of magnitude greater than that characteristic of ordinary boiling. Note the bulging of the bubble surface into the host fluid caused by the reaction to the thrust of the jet. This photograph dramatically demonstrates why vapor explosions are so violent.

After vaporization is completed, excess over pressure within the bubble causes it to expand rapidly and subsequently collapse and oscillate. The bubble in Fig. 4 was photographed during the first rebound following the initial collapse of the bubble. The breakup of the spherical bubble is a result of the Rayleigh-Taylor instability which is driven by the large accelerations of the bubble surface. Further information on the dynamics of the explosive boiling process can be found in references 1 and 2.

Keywords
vapor bubble; Rayleigh-Taylor instability

[1] D. L. Frost, "Dynamics of explosive boiling of a droplet," *Phys. Fluids* **31** (9) 2554–2561 (1988).

[2] D. L. Frost, "Initiation of explosive boiling of a droplet with a shock wave," *Experiments in Fluids* **8**, 121–128, (1989).

Figure 1

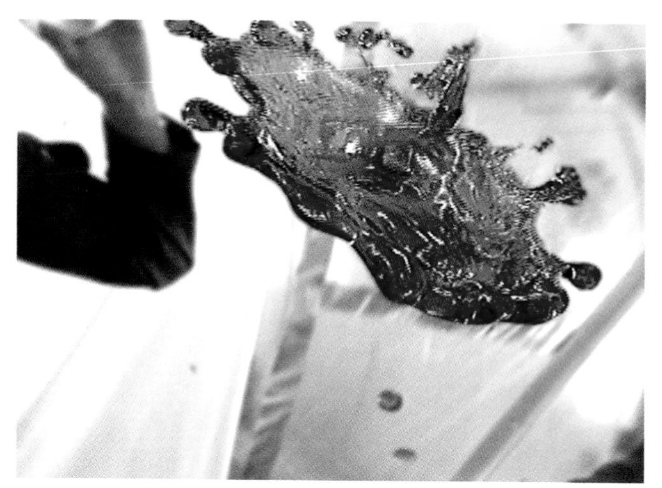

Figure 3

Figure 2

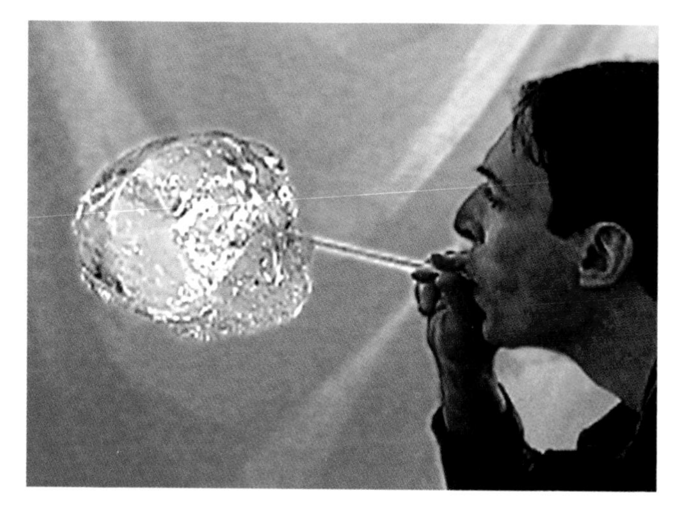

Figure 4

Water balloon rupture in low-g

M. M. Weislogel

NASA Lewis Research Center

and S. Lichter

Northwestern University

A qualitative study of the bursting of water balloons in a simulated low-gravity environment was conducted aboard NASA Lewis's DC-9 aircraft. Following rupture by a syringe needle, the balloon retracts tangent to the water surface leaving a smooth surface near the puncture location, Fig. 1. Asymmetries soon develop as the membrane rips apart, ejecting a directional spray from the surface. When the balloon parts from the drop entirely, it causes a large deformation of the remaining liquid mass leading to significantly underdamped oscillations which persist for the duration of the simulation.

In Fig. 2 an approximately 2.5 liter blob of undulating water hovers after rupture of the membrane. A free floating, red-dyed water blob flattens, crowns, and breaks up after impact by an impinging blue-dyed water jet, Fig. 3. In Fig. 4 a large air bubble is blown into a free floating blob using a straw. Such flows are *unearthly* in that a balance is struck between capillary and inertial forces over truly large length scales.

The tests were performed to develop techniques to rapidly deploy large liquid drops in a microgravity environment. The footage has also proven of general interest and is used to introduce students to low-g phenomena. Quicktime movies may be found at http://zeta.lerc.nasa.gov/balloon/blob.htm. Related ground tests were also performed using high-speed video photography and may be found at http://zeta.lerc.nasa.gov/balloon/hs.htm.

Keywords

capillary wave; inertial oscillations; capillary length-scale.

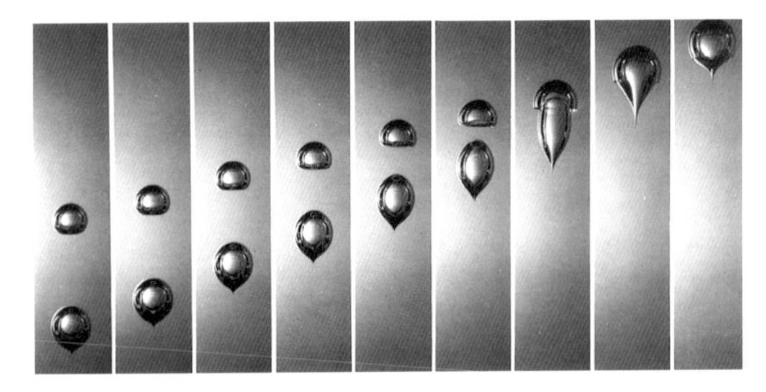

Figure 1

Figure 2

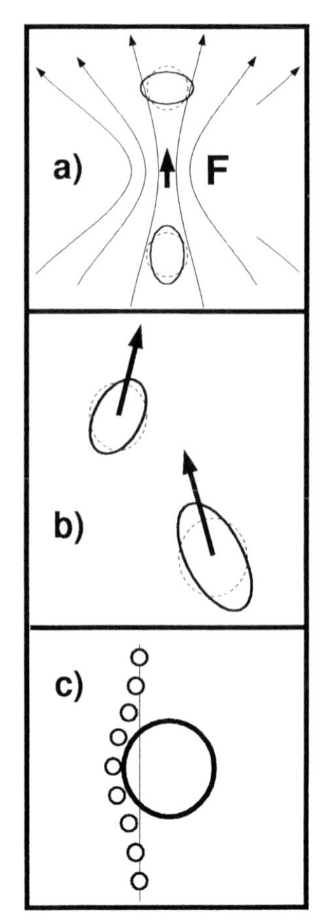

Figure 3

Interactions between bubbles at low Reynolds numbers

Michael Manga and H. A. Stone

Harvard University

In Fig. 1 we show a sequence of photographs of a large air bubble following a small air bubble in a large container of corn syrup. For this axisymmetric configuration, the small bubble spreads over the surface of the larger bubble. The experiment in Fig. 2 illustrates the alignment of initially horizontally offset bubbles. In this case, interaction-induced deformation of the bubbles leads to lateral components of translation and the eventual alignment of the bubbles. Photographs in Figs. 1 and 2 are taken at 5 sec intervals. The Bond numbers for the experiments are large, $\mathcal{B} = \Delta\rho g a^2/\sigma \approx 50$, and the Reynolds numbers are small $\mathcal{R} = \rho U a/\mu \approx 10^{-3}$. The Bond number represents the ratio of buoyancy forces to interfacial tension forces. Here, σ is interfacial tension, g is gravitational acceleration, a is the radius of the larger bubble, $\Delta\rho$ is the density defect, μ is the fluid viscosity, and U the bubble rise speed.

Since streamlines in front of a rising bubble diverge, the leading bubble is flattened (oblate distortion); conversely, the trailing bubble is extended (prolate distortion) due to the convergence of streamlines produced by the leading bubble [see Fig. 3(a), in which a point force **F** is used to represent the flow created by a translating bubble]. In Figs. 3(a) and 3(b), dashed lines denote the undeformed bubble shapes, and solid lines denote the deformed shapes. The alignment demonstrated in Fig. 2 occurs since the effect of bubble interactions is to deform the trailing bubble into a prolate shape, suitably inclined with respect to the vertical; the leading bubble is deformed into an oblate shape. If deformation is sufficiently large, the resulting translational velocities may favor alignment, as indicated in Fig. 3(b).

The corresponding problem of two translating spherical bubbles evolves so that, relative to a reference frame translating with the larger sphere, the small sphere is swept toward the back, as shown in Fig. 3(c). The off-axis configuration for spherical bubbles is unstable as a consequence of the hyperbolic stagnation point at the front of the larger bubble. A comparison of Fig. 3(c) with the experiments shown in Figs. 1 and 2, demonstrates the importance of deformation on bubble dynamics. Deformation-induced interactions typically result in alignment and the eventual coalescence of bubbles.

We acknowledge the support of NSF, ACS, and IGPP grants and thank Harvard undergraduates Karin Riley and Joe Rice for assistance with the experiments.

Keywords

bubble collision; air bubbles.

Figure 1

Falling Drop

Air

Water

Rising Bubble

Figure 2

Bursting soap bubble

James Buchholz, Lorenz Sigurdson, and Bill Peck

University of Alberta

This vortex ring was generated by a smoke-filled bursting soap bubble resting on a free surface. This is similar to the simpler case of a rising bubble penetrating a free surface, the "flip" experiment of a falling water drop impacting a free surface[1,2]. Comparing these two rings may help understand the vorticity generation mechanism at an air-water interface. This is a triple exposure, first by the spark used to break the bubble, once by a strobe very shortly after the bubble has been pierced by the spark, and once at a later time showing the resulting vortex ring (base width of bubble = 1.54 cm). More details can be found in reference 3.

The authors thank Mr. Bernie Faulkner for his technical assistance and the Natural Sciences and Engineering Research Council of Canada for their support (Grant No. OGP41747).

Keywords

triple exposure; vortex ring.

[1] B. Peck and L. W. Sigurdson, *Phys. Fluids* A **3**, 2032 (1994).

[2] B. Peck and L. W. Sigurdson, "The three-dimensional vortex structure of an impacting water drop," *Phys Fluids* **6**, 564–576 (1994).

[3] J. Buchholz and L. W. Sigurdson, "The kinematics of the vortex ring structure generated by a bursting bubble," *Phys. Fluids*, **12**, 42–53 (2000).

5 Complex fluids

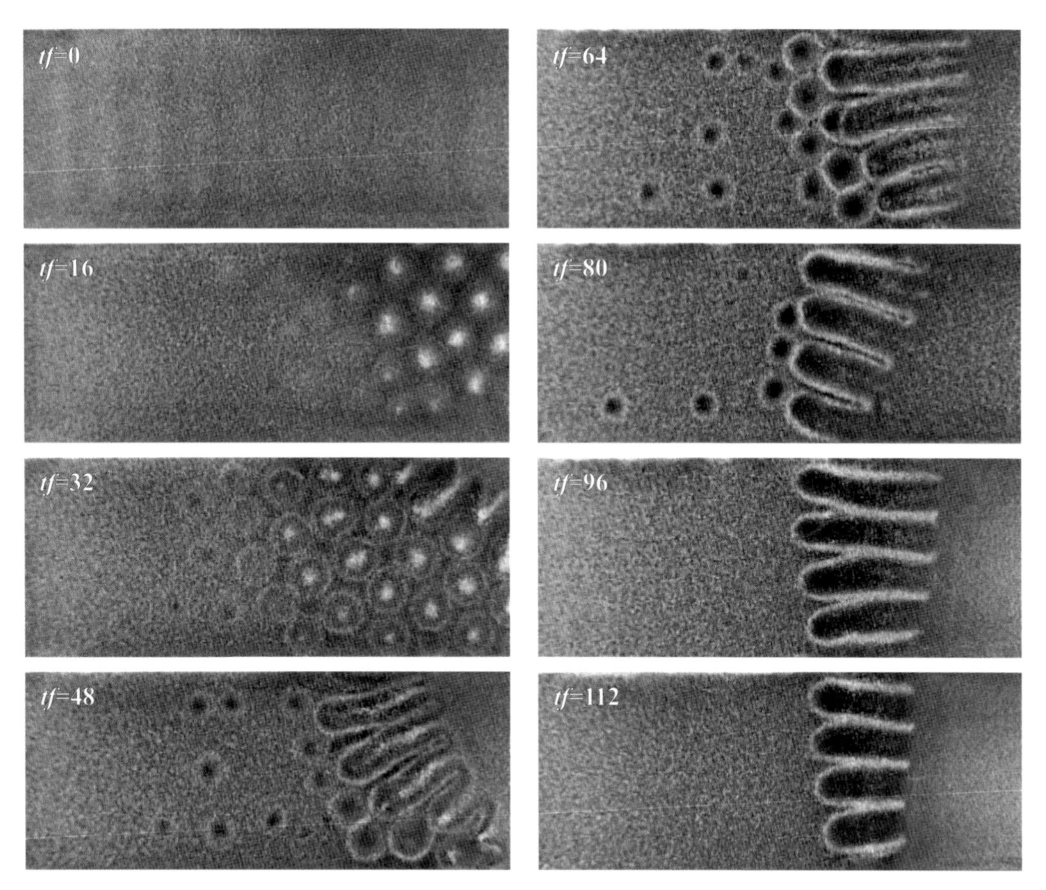

Figure 1

Interface motion in a vibrated granular layer

I. Aranson and D. Blair

Argonne National Laboratory

Peter Vorobieff

Los Alamos National Laboratory

Granular materials are now recognized as a distinct state of matter, and studies of their behavior form a fascinating interdisciplinary branch of science. The intrinsic dissipative nature of the interactions between the constituent macroscopic particles gives rise to several basic properties specific to granular substances, setting granular matter apart from the conventional gaseous, liquid, or solid states.[1]

Thin layers of granular materials subjected to vertical vibration exhibit a diversity of patterns.[2] The particular pattern is determined by the interplay between driving frequency f and the acceleration amplitude Γ. Interfaces in vibrated granular layers, existing for large enough amplitude of vibration, separate large domains of flat layers oscillating with opposite phase. These two phases are related to the period-doubling character of the flat layer motion at large plate acceleration. Interfaces are either smooth or "decorated" by periodic undulations depending on parameters of vibration.[3] An additional subharmonic driving results in a controlled displacement of the interface with respect to the center of the experimental cell.[4] The speed and the direction of the interface motion are sensitive to the phase and amplitude of the subharmonic driving.

The image sequence above shows interface nucleation and propagation towards the center of the cell, with dimensionless time tf labeled in each image. The interface forms at the right side wall of the cell due to small-amplitude phase-shifted subharmonic driving. After the additional driving stops, the interface moves towards the center, creating small-scale localized structures in the process.

Keywords

Faraday forcing; vibrating plate; subharmonic forcing.

[1] H. M. Jaeger, S. R. Nagel, and R. P. Behringer, *Phys. Today* **49**, 32 (1996); *Rev. Mod. Phys.* **68**, 1259 (1996).

[2] F. Melo, P. B. Umbanhowar, and H. L. Swinney, *Phys. Rev. Lett.* **72**, 172 (1994); **75**, 3838 (1995); P. B. Umbanhowar, F. Melo, and H. L. Swinney, *Nature* **382**, 793 (1996); *Physica* **A 249**, 1 (1998).

[3] P. K. Das and D. Blair, *Phys. Lett.* **A 242**, 326 (1998).

[4] I. Aranson, D. Blair, W. Kwok, G. Karapetrov, U. Welp, G. W. Crabtree, V. M. Vinokur, and L. Tsimring, "Controlled dynamics of interfaces in a vibrated granular layer," *Phys. Rev. Lett.* **82**, 731 (1999).

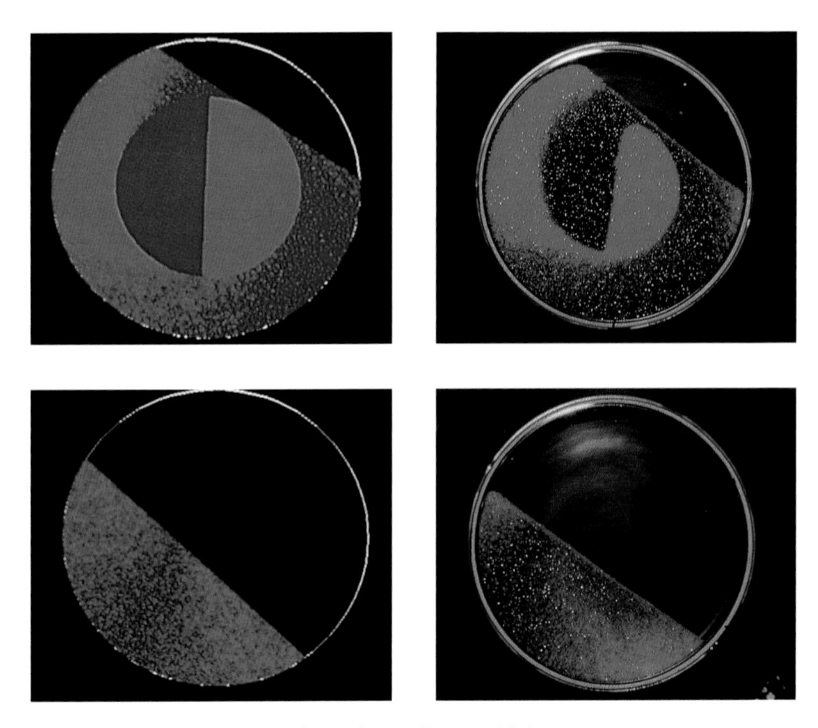

Figure 1 Computations (left) and experiments (right).

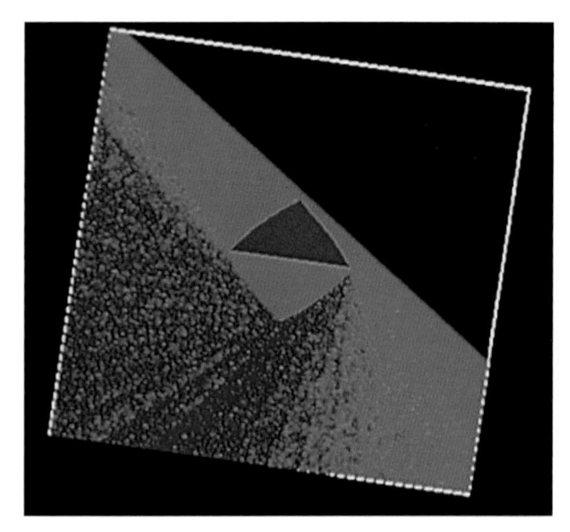

Figure 2 Computations (top) and experiments (bottom).

Mixing patterns for surface dominated granular flows

G. Metcalfe, T. Shinbrot, J. J. McCarthy, and J. M. Ottino *Northwestern University*

Avalanches mix flowing granular solids. The photographs show experiments and computations of granular mixing in slowly rotating containers. Mixing occurs through the action of successive avalanches across the surface. Material below the surface rotates as a solid body with the container.[1]

Experimentally, the disk is thin enough that the dynamics occur in a plane. The particles are dyed salt cubes – mean diameter 0.6 mm – and the disk is 240 grains in diameter and 40 grains deep.

Geometric modeling successfully captures the mixing behavior. The mixing behavior of powders can be divided into a geometric part and a dynamical part, and for monodisperse and weakly cohesive particles the geometric aspects dominate.

Figure 1 shows mixing patterns from simulation (left) and experiment (right) after two disk revolutions. For below half full (bottom), there is a similar color pattern and degree of mixedness. For above half full (top), an unmixed core appears. The geometric modeling applies to any container shape; Fig. 2 shows results in a square after 1–1/4 revolutions.

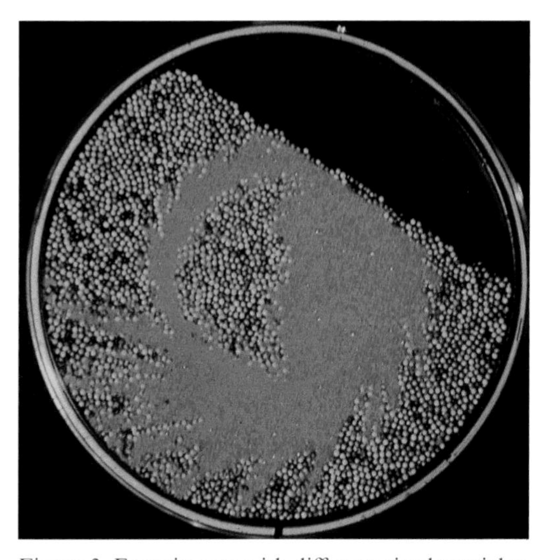

Figure 3 Experiments with different-sized particles.

When the particles are different, e.g., in size or shape, more complicated behavior can occur and particles can even unmix. Figure 3 shows one of the more complicated patterns in a mixture of the salt cubes and 1.8 mm sugar balls.

Keywords

salt cubes; particle mixing; granular mixing.

[1] G. Metcalfe, T. Shinbrot, J. J. McCarthy, and J. M. Ottino, "Avalanche mixing of granular solids," *Nature* **374**, 39 (1995).

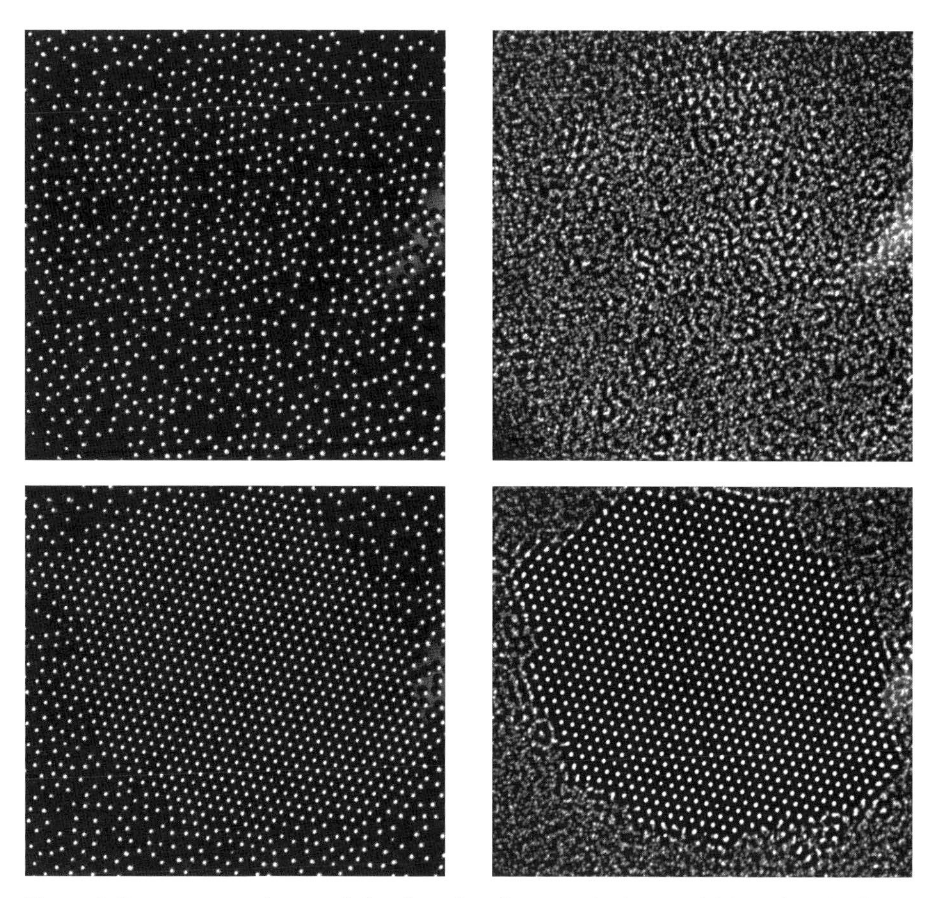

Figure 1 Instantaneous images (left column) and composite images (right column) of a 2D driven granular layer as viewed from above. Upper left: Instantaneous image when $\Gamma = 0.78$. Upper right: Bright peaks in the composite image at $\Gamma = 0.78$ denote clusters. Lower left: Instantaneous image of collapse at $\Gamma = 0.76$. Lower right: Composite image of the collapse demonstrates the two-phase behavior.

Crystallization in a 2D granular fluid

J. S. Olafsen and J. S. Urbach

Georgetown University

Granular media can be fluidized by imparting energy, for example by vibration. The resulting fluctuating motion of the grains displays many similarities with the random motion of molecules in liquids and gases, but with some striking differences due to the dissipation that occurs during inter-particle collisions. Instantaneous and composite images demonstrate surprising collective behavior in an experiment comprised of 8000 uniform spherical stainless steel particles on a vertically shaken plate at a frequency of 70 Hz. (The dimensionless acceleration is defined as $\Gamma = A(2\pi\nu)^2/g$, where g is the acceleration due to gravity and A is the amplitude of the plate displacement.)

A camera positioned above the plate captures the nearly 2D behavior of the system. A diffuse light source illuminates an area of approximately one half of the upper surface of each particle. The layer is initially fluidized at $\Gamma = 1$. By slowly lowering the acceleration, the medium is cooled while remaining near the steady state. At $\Gamma = 0.78$, non-uniform density distributions in an instantaneous image (upper left) and non-uniform pixel intensities in a 1 second composite image (upper right) appear as a consequence of clustering of the particles, despite the absence of any attractive forces. A cluster then becomes the nucleation point for a "collapse" when the acceleration is decreased to $\Gamma = 0.76$ (lower left). The collapse is the ordered condensate of particles that are in contact with each other and the plate and remain motionless while the surrounding particles continue in a gas-like phase at reduced density. This is demonstrated in a 1 second composite image (lower right). The two-phase behavior will persist as long as the driving is maintained. To refluidize the condensate, the plate's acceleration must be increased slightly to $\Gamma = 0.80$.

This work was supported by Research Corporation and the Petroleum Research Fund.

Keywords

steel balls; vibrating plate; clustering.

Figure 1 Jet formation upon impact of a steel ball onto loose, very fine sand. A splash is formed (b), then a jet emerges (c) and (d). After reaching its highest point, clusters are formed through inelastic collisions (e). A granular eruption concludes the series of events (f).

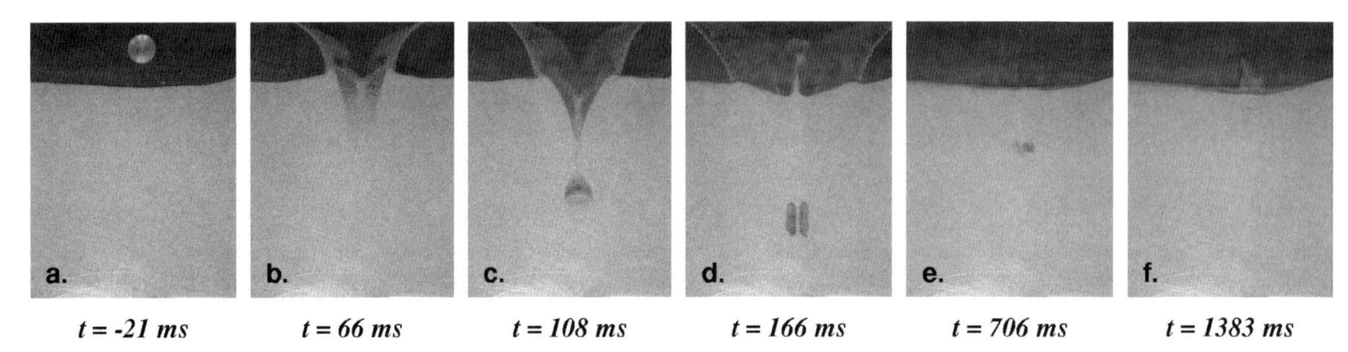

Figure 2 The two-dimensional setup reveals the collapse of the void and the formation of the jet: The cylinder creates a void in the sand (b). Upon closure an air bubble is entrained (c) and two jets are produced: one upward, the other downward (d). The bubble slowly rises through the sand (e) and upon reaching the surface causes the granular eruption.

Granular eruptions: void collapse and jet formation

René Mikkelsen, Michel Versluis, Elmer Koene, Gert-Wim Bruggert, Devaraj van der Meer, Ko van der Weele, and Detlef Lohse

University of Twente

A steel ball dropped onto loose, very fine sand creates a jet.[1] Here, a container is filled with dry and with an average grain size of 40 μm. Air is blown through the sand through a sintered bottom plate in order to make it loose and homogeneous. The air is slowly turned off before the experiments. A steel ball (25 mm diameter) is dropped into the sand [Fig. 1(a)] and the dynamics of the sand is recorded with a digital high-seed camera at 1000 frames per second.

The series of events is as follows:[2]

1. **Splash:** The ball vanishes in the sand and a crown-like splash is created. Inhomogeneities develop in the crown, presumably due to the inelastic particle–particle interaction [Fig. 1(b)].
2. **Void collapse and jet formation:** The ball creates a cylindrical void which collapses because of the sand pressure. When the accelerated sand grains from the sidewalls of the void collide on the axis of the void, a jet is formed [Figs. 1(c) and 1(d)]. In all our experiments the jet height exceeded the release height of the ball. When it comes down again the jet breaks up due to inelastic collisions of the sand particles, forming granular clusters [Fig. 1(e)].
3. **Eruption:** An air bubble is formed in the void and slowly rises in the sand. When it hits the surface, it causes a granular eruption, resembling a volcano [Fig. 1(f)].

A two-dimensional setup reveals the jet-formation process. The ball was replaced by a cylinder which was dropped into a bed of sand between two transparent plates. The jet takes the form of a sheet making it less intense. The void formation [Fig. 2(b)] and collapse [Fig. 2(c)] are observed through the sidewalls. As the void closes an air bubble is entrained and it is seen [Fig. 2(d)] that, in fact, *two* jets are formed, one going upward and one going downward.

Keywords

particle splash; particle jetting; bubble eruption.

[1] S. T. Thoroddsen and A. Q. Shen, *Phys. Fluids* **13**, 4 (2001).

[2] See fdweb.tn.utwente.nl/granular/granular_eruptions.mpg for a video showing the series of events.

(a)

(b)

Figure 1 A well-mixed 10% suspension (a) before shearing and (b) after shearing, with the inner cylinder rotating at 9 rpm when the Couette is filled up to 95% of the available gap volume.

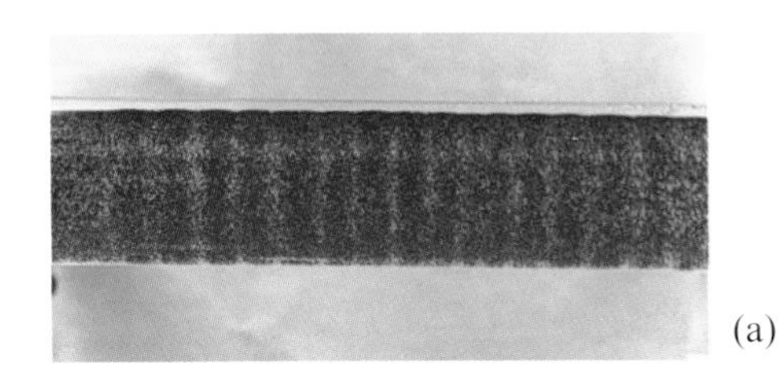
(a)

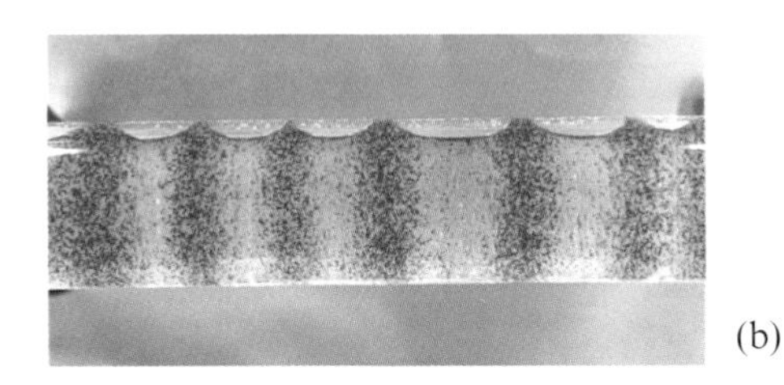
(b)

Figure 2 A 15% suspension at the 95% fill level is sheared at (a) 2.5 rpm and (b) 9 rpm.

(a)

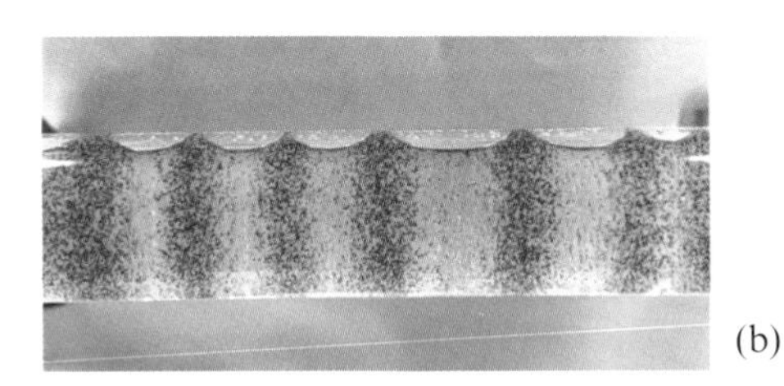
(b)

Figure 3 Segregation and band formation for (a) a 5% suspension at 9 rpm and a 90% fill level and (b) a 15% suspension at 9 rpm and a 95% fill level.

(a)

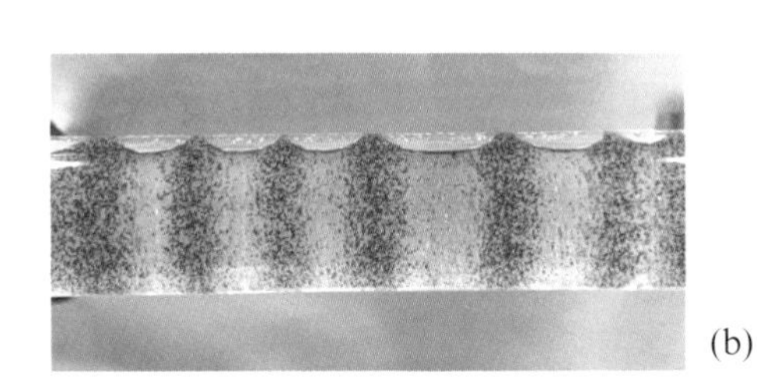
(b)

Figure 4 Segregation for (a) a 15% suspension at a 50% fill level and (b) a 15% suspension at a 95% fill level, sheared at 9 rpm.

Particle segregation in monodisperse sheared suspension

Mahesh Tirumkudulu, Anubhav Tripathi, and Andreas Acrivos

The Levich Institute, The City College of the City University of New York, New York

The photographs shown above reveal startling evidence of a new instability[1] in suspensions of monodisperse neutrally buoyant spherical particles in a Newtonian liquid medium being sheared in a partially filled horizontal Couette device. Upon being sheared, the suspension separates itself into alternating regions of high and low particle concentration along the length of the tube. The horizontal Couette device consisted of two concentric cylinders of radius 1.21 cm and 1.90 cm. The outer cylinder was kept fixed and the inner cylinder was free to rotate. The suspending liquid, prepared from a combination of Triton X-100, $ZnCl_2$ and water, was Newtonian and had a viscosity of 3400 cP (at 23°C). The dispersed phase consisted of spherical acrylic particles (colored red for visibility) of mean diameter 462.5 μm and had a density equal to that of the suspending liquid (1.174 gm/cc).

Shown in Figs. 1(a) and 1(b) are the photographs of a uniformly mixed suspension before and after it had been sheared at 9 rpm. However, no segregation was observed when the Couette was completely filled. Figures 2(a) and 2(b) show the effect of rotation rate on segregation and band formation on a 15% suspension. At the higher rotation rate the number of bands has decreased but the degree of segregation has been enhanced. The effect of variation in total particle concentration can be seen in Figs. 3 (a) and 3(b). When the fill level for the 15% suspension was varied [Figs. 4(a) and 4(b)], the bands were smaller in width and less distinct for the 50% fill level than for the 95% fill level. It is to be noted that such an instability does not arise for pure fluids of comparable viscosity.

The work described in this paper has been supported in part by a grant from the DOE DE-FG02-90ER14139.

[1] M. Tirumkudulu, A. Tripathi, and A. Acrivos, "Particle segregation in monodisperse sheared suspensions," *Phys. Fluids* **11**, 507 (1999).

Keywords

Couette device; shear-induced segregation.

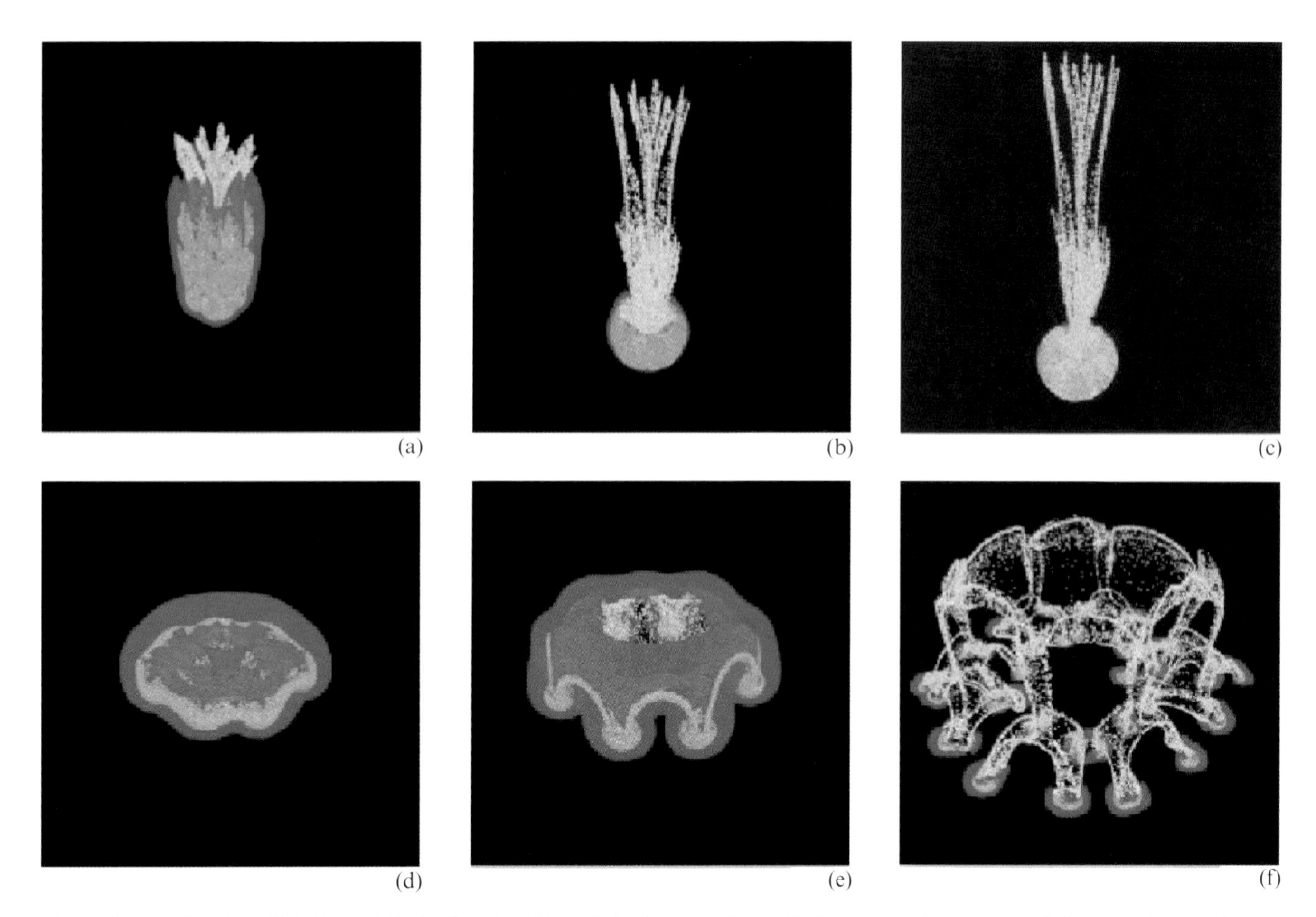

Figure 1 Visualization of fluid vorticity (red) and solid particles (white) of an initially spherical suspension falling due to gravity. Case A: (a)–(c); Case B: (d)–(f).

Simulation of particle laden flows using particle methods

J. H. Walther ETH Zürich

S.-S. Lee Stanford University

P. Koumoutsakos ETH Zürich and NASA Ames

We present simulations of an initially spherical suspension of solid particles falling due to gravity in a viscous incompressible fluid. The numerical simulations are performed using three-dimensional viscous, vortex methods[1] with a *two-way* coupling between the fluid and the particles.[2]

We consider two cases: (A) the suspension is falling in a nonzero initial vorticity field, generated by allowing, for a short time, only one-way viscous coupling and (B) the suspension is inserted in a fluid with no initial vorticity field.

In case A [Figs. 1(a)–(c)], the particle suspension is entrapped by the initial vorticity field which dominates the dynamics of the flow. The solid particles are wrapped around the core of the vortex ring and are shed in its wake. This results in the formation of "Tintenpilze."[3] In case B [Figs. 1(d)–(f)], the falling of the solid particles imparts vorticity on the fluid elements, forming a vortex ring. The vortex ring becomes unstable and induces a bifurcation of the suspension of particles which in turn creates new vortex rings in a cascade-like fashion. The phenomena are qualitatively similar to experimental observations of liquid drops injected in lighter fluids.[4]

Keywords

simulation; vortex ring; Tintenpilze.

[1] G.-H. Cottet and P. Koumoutsakos, *Vortex Methods: Theory and Practice* (Cambridge University Press, New York, 2000).

[2] J. H. Walther and P. Koumoutsakos, "Three-Dimensional Vortex Methods for Particle-Laden Flows with Two-Way Coupling," *J. Comput. Phys.* **167**, 39–71 (2001).

[3] D. W. Thompson, *On Growth and Form* (Dover, New York, 1992).

[4] J. J. Thomson and H. F. Newall, *Proc. R. Soc.* **39**, 417 (1885).

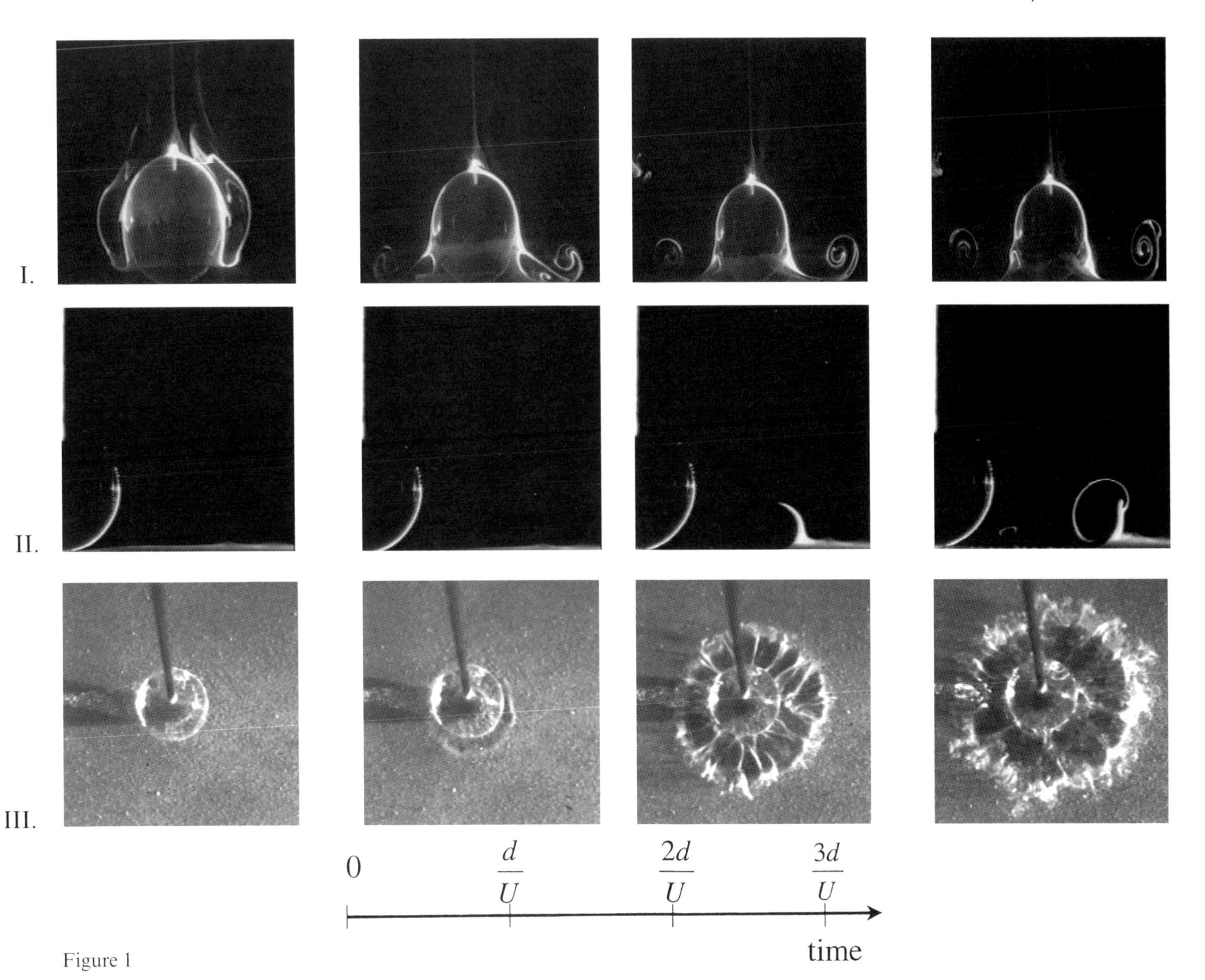

Figure 1

Resuspension by an impacting sphere

I. Eames *Bristol University*

S. B. Dalziel *Cambridge University*

We report on a series of experiments which show how the flow generated by a rigid sphere moving toward a wall is capable of resuspending dust. The sphere is mounted on a rigid rod, driven with a constant velocity toward the wall, and stops when contact is made between the wall and sphere. The flow is characterized by a Reynolds number $Re = dU/\nu$ based on the sphere diameter d, impact velocity U and kinematic viscosity ν of the ambient fluid. The wake vortex behind a translating sphere plays an important role in resuspending dust, and sequence I ($Re = 850$) shows the dynamics of the wake (visualized by a light sheet where the wake is rendered visible by the precipitation of tin chloride) following the sphere impact. The first image in sequence I corresponds to the sphere making contact with the wall, and the time difference between subsequent images is d/U. The wake vortex generates a secondary vortex as it passes over the surface of the sphere. The fluid adjacent to the wall is visualized in sequence II ($Re = 850$) by introducing a thin layer of fluorescein dye, and this sequence serves to illustrate how the wake vortex pushes fluid initially adjacent to the wall to one side, which is entrained by the wake vortex, and transported away from the wall. Many aspects of this flow may be understood from studies of vortex impact on walls.[1] Sequence III ($Re = 3100$) shows the resuspension a thin layer of dust by the wake vortex as it threads over the sphere and strikes the wall. This sequence shows the azimuthal instability of the secondary vortex which collects the dust into piles resembling the spokes on a wheel. A more comprehensive description of these flows is given by Eames and Dalziel.[2]

I.E. gratefully acknowledges financial support through the E.P.S.R.C. Overseas Travel Grant (GR/M45856) to attend the 1998 APS/DFD meeting.

[1] J. D. A. Walker, C. R. Smith, A. W. Cerra, and T. L. Doligalski, "The impact of a vortex ring on a wall," *J. Fluid Mech.* **181**, 99 (1987).

[2] I. Eames and S. B. Dalziel, "Dust resuspension by the flow around an impacting sphere," *J. Fluid Mech.* **403**, 305–328 (2000).

Keywords
dust resuspension; wake vortex.

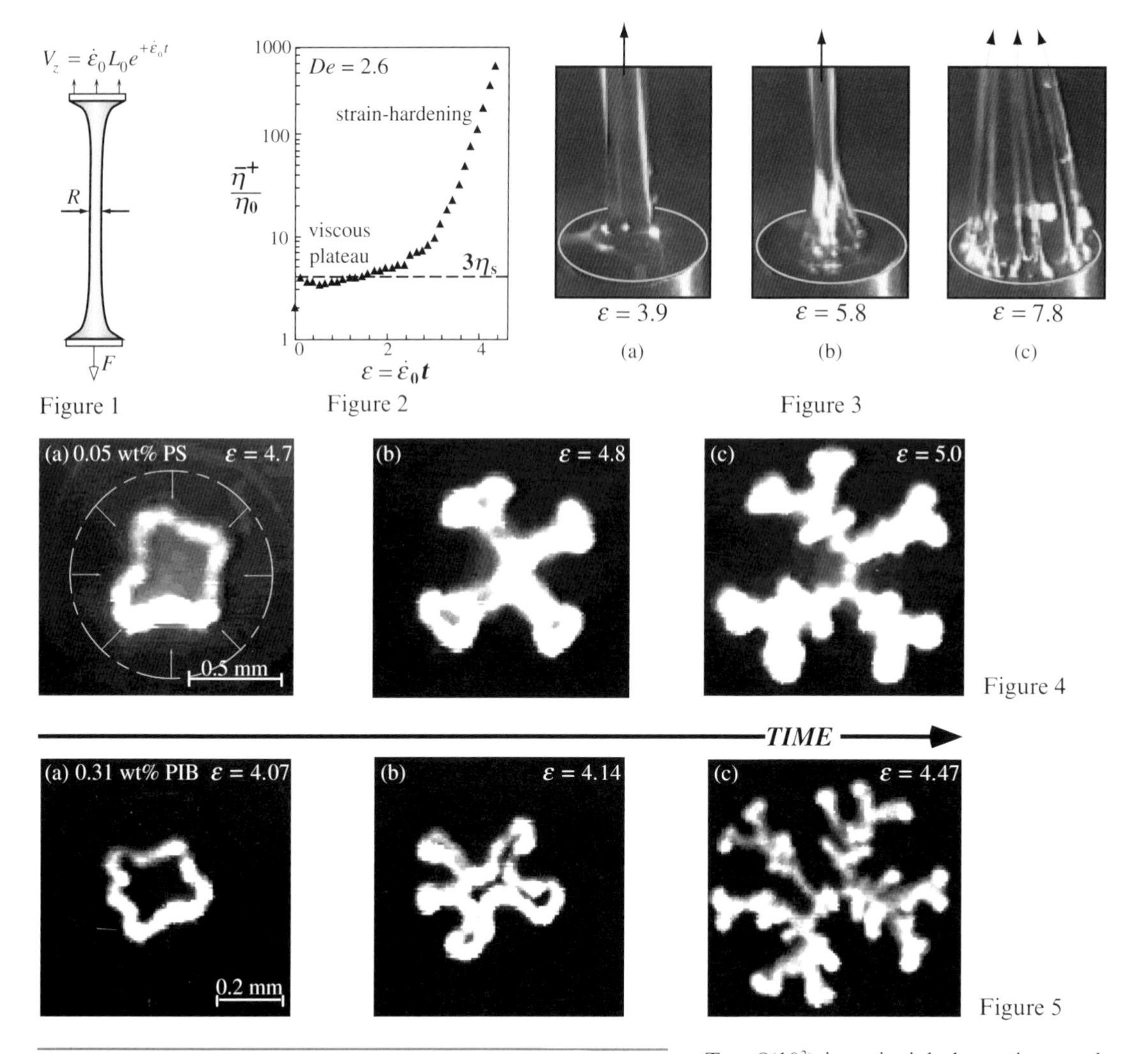

Figure 1

Figure 2

Figure 3 (a) (b) (c)

Figure 4

Figure 5

Elastic instability in elongating fluid filaments

Shelley L. Anna, Stephen H. Spiegelberg, and Gareth H. McKinley

Harvard University

To quantify the response of dilute polymer solutions in strong extensional flows, we measure the transient extensional viscosity by exponentially elongating cylindrical fluid filaments as indicated in Fig. 1. Measurement of the filament radius $R(t)$ and the tensile force $F(t)$ exerted on the endplate (corrected for surface tension σ) leads to the definition of a Trouton ratio $\mathrm{Tr} = \bar{\eta}^+/\eta_0 = F/(\pi R^2 \dot{\varepsilon}_0) - \sigma/\dot{\varepsilon}R$.[1] Ideal elastic polymer solutions (or "Boger fluids") exhibit pronounced strain hardening and Trouton ratios, $\mathrm{Tr} \approx O(10^3)$ in uniaxial elongation as shown in Fig. 2. At high strains, this rapid tensile stress growth leads to an unexpected elastic free-surface instability.[2] The cylindrical fluid filament loses axisymmetry and an azimuthally periodic series of thin elastic fibrils evolves near the stationary endplate as shown in Fig. 3. As the strain and tensile stress increase, the filament can ultimately separate entirely from the rigid endplate. By using a glass endplate, plan views of the evolution in spatial structure of the instability are obtained, as shown in Figs. 4 and 5. The disturbance at the radially contracting free surface of the filament is initially periodic and characterized by an azimuthal mode number that varies with fluid composition, Deborah number, and strain. As the straining continues, thin finger-like lobes form which grow *radially outward*, tip-splitting and branching in a familiar fractal-like pattern. This elastically driven instability is cohesive rather than adhesive in nature and can be characterized as a modified Taylor meniscus instability[3] in which destabilizing elastic normal stresses in the bulk of the fluid compete with the stabilizing interfacial tension of the curved free surface.

This research is supported by NASA under Grant No. NAG3-1385. S. L. A. thanks the Fannie and John Hertz Foundation for supporting her graduate research. More information and Quicktime movies of these observations can be found at the URL http://stokes.harvard.edu/.

[1] S. H. Spiegelberg, D. C. Ables, and G. H. McKinley, "The role of end-effects on measurements of extensional viscosity in filament stretching rheometers," *J. Non-Newtonian Fluid Mech.* **64**, 229 (1996).

[2] S. H. Spiegelberg and G. H. McKinley, "Stress relaxation and elastic decohesion of viscoelastic polymer solutions in extensional flow," *J. Non-Newtonian Fluid Mech.* **67**, 49 (1996).

[3] A. D. McEwan and G. I. Taylor, "The peeling of a flexible strip attached by a viscous adhesive," *J. Fluid Mech.* **26**, 1 (1966).

Keywords
Boger fluid; extensional flow; fluid filament.

6 Flows with interfaces

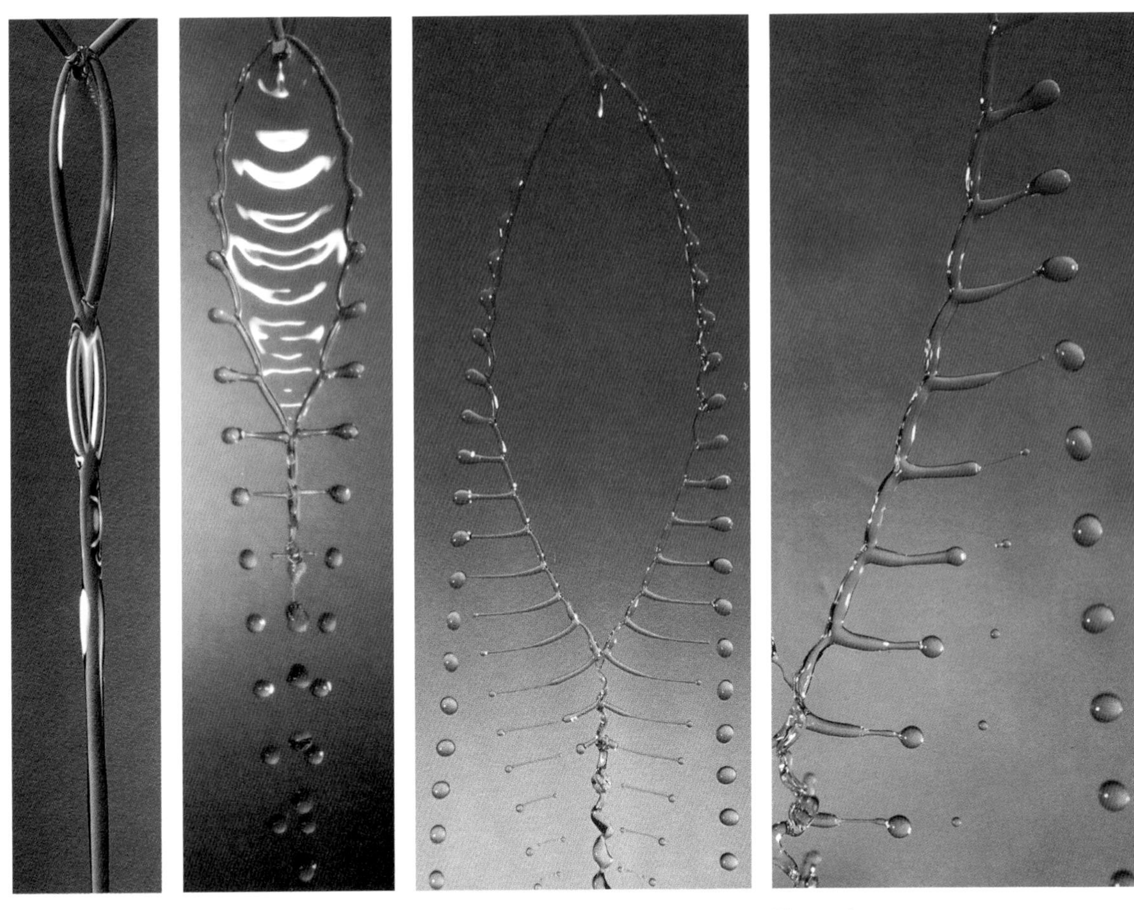

Figure 1 Figure 2 Figure 3 Figure 4

Fluid fishbones

Alexander E. Hasha and John W. M. Bush

Massachusetts Institute of Technology

We examine the form of the free surface flows resulting from the collision of equal jets at an oblique angle. Glycerol-water solutions with viscosities of 15–50 cS were pumped at flow rates of 10–40 cc/s through circular outlets with diameter 2 mm. Characteristic flow speeds are 1–3 m/s. Figures 2–4 were obtained through strobe illumination at frequencies in the range 2.5–10 kHz.

At low flow rates, the resulting stream takes the form of a steady fluid chain,[1] a succession of mutually orthogonal fluid links, each comprised of a thin oval sheet bound by relatively thick fluid rims (Fig. 1). The influence of viscosity serves to decrease the size of successive links, and the chain ultimately coalesces into a cylindrical stream.

As the flow rate is increased, waves are excited on the sheet, and the fluid rims become unstable (Figs. 2 and 3). Droplets form from the sheet rims but remain attached to the fluid sheet by tendrils of fluid that thin and eventually break. The resulting flow takes the form of fluid fishbones, with the fluid sheet being the fish head and the tendrils its bones. Increasing the flow rate serves to broaden the fishbones.

In the wake of the fluid fish, a regular array of drops obtains, the number and spacing of which is determined by the pinch-off of the fishbones (Fig. 4). At the highest flow rates examined, the flow is reminiscent of that arising in acoustically excited fan-spray nozzles.[2]

[1] M. F. G. Johnson, M. J. Mikisis, R. A. Schluter, and S. G. Bankoff, "Fluid chains produced by obliquely intersecting viscous jets connected by a thin free liquid film," *Phys. Fluids* **8**, S2 (1996).

[2] For example, see N. Dombrowski's photo in *An Album of Fluid Motion*, ed. M. Van Dyke (Parabolic, Stanford, 1988), p. 87.

Keywords

spray generation; liquid chain; ligament breakup.

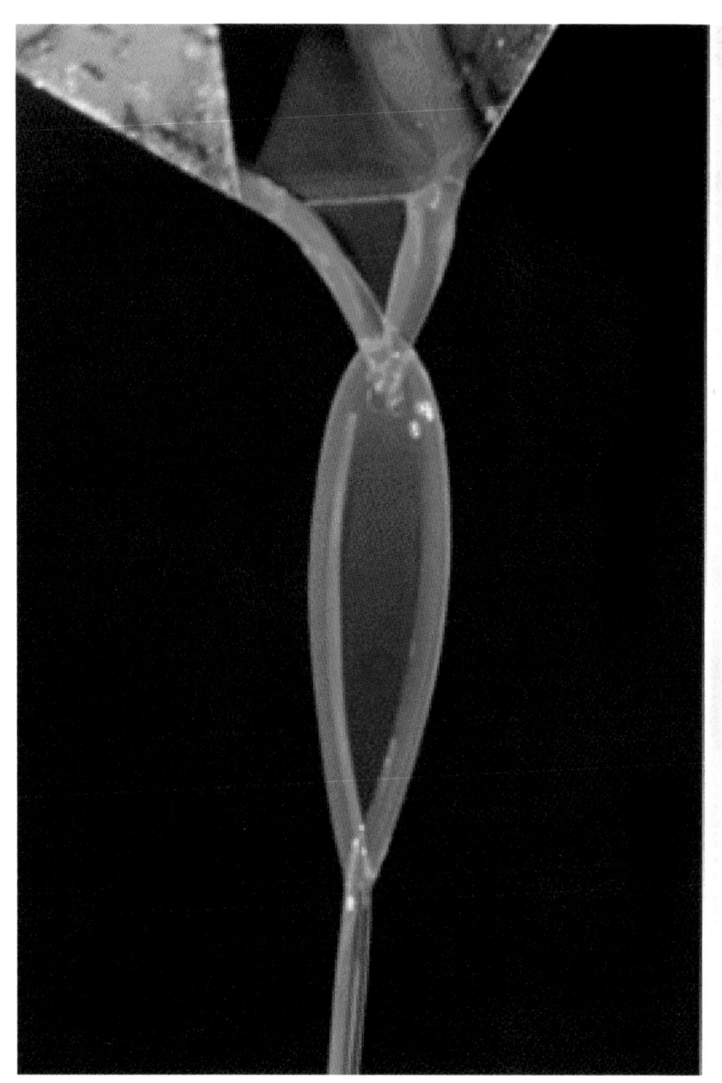
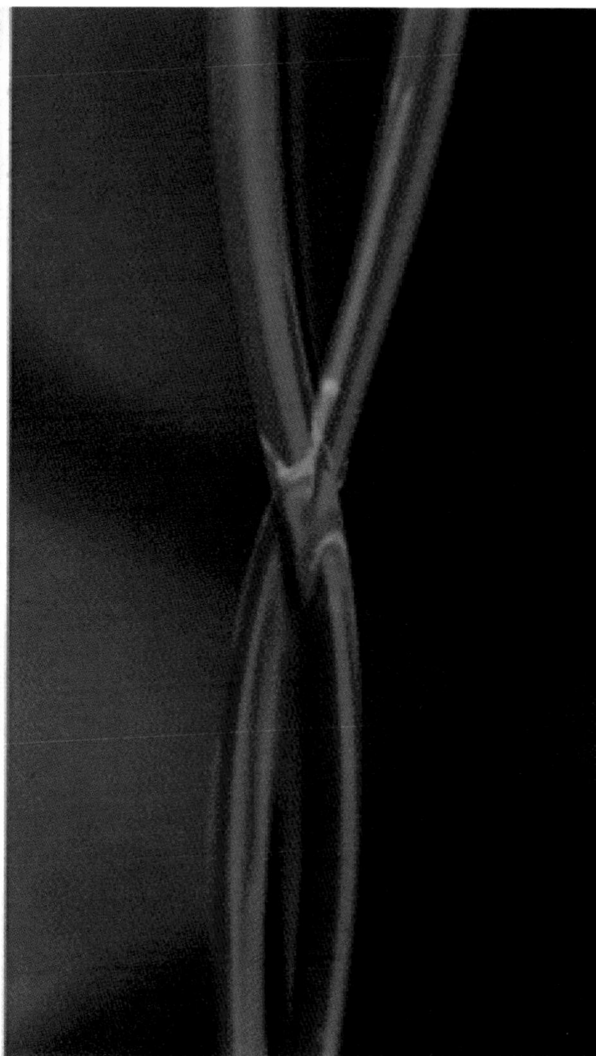

Figure 1

Fluid chains produced by obliquely intersecting viscous jets connected by a thin free liquid film

M. F. G. Johnson, M. J. Miksis, R. A. Schluter, and S. G. Bankoff

Northwestern University

A rather unique fluid dynamic configuration was observed in the course of studying rivulet formation from a contact line moving down an inclined plate about 1 m^2. The exiting liquid was funneled via an inclined flat trough with narrowing sidewalls into an open vessel below. The liquid flowed principally along the funnel sidewalls, but also as a thin sheet in between. This formed two obliquely colliding falling jets with a thin liquid film connecting them. To our surprise the jets retained their character for long distances downstream, colliding and recolliding to form stable liquid rings, always with the surface tension of the liquid sheet pulling the outwards-pointing rebound jets back into the ring configuration. In each case, the rebound jets conserved momentum normal to the plane of entering jets by rotating the plane of exiting jets 90°. The jet and sheet structure is made fluorescent by illuminating the liquid, consisting of a 50% by volume mixture of glycerin and water with small amounts of fluorescein, by strong ultraviolet light. The entering jets are about the same diameter, and their centerlines appear close to intersecting. Thus, a mixing region is observed from which the exiting jets emerge.

Thanks are due to Leslie Hocking for his observations in our laboratory. This work is supported by the National Science Foundation, U. S. Department of Energy, and NASA.

Keywords

fluorescein; jet collisions; liquid chain.

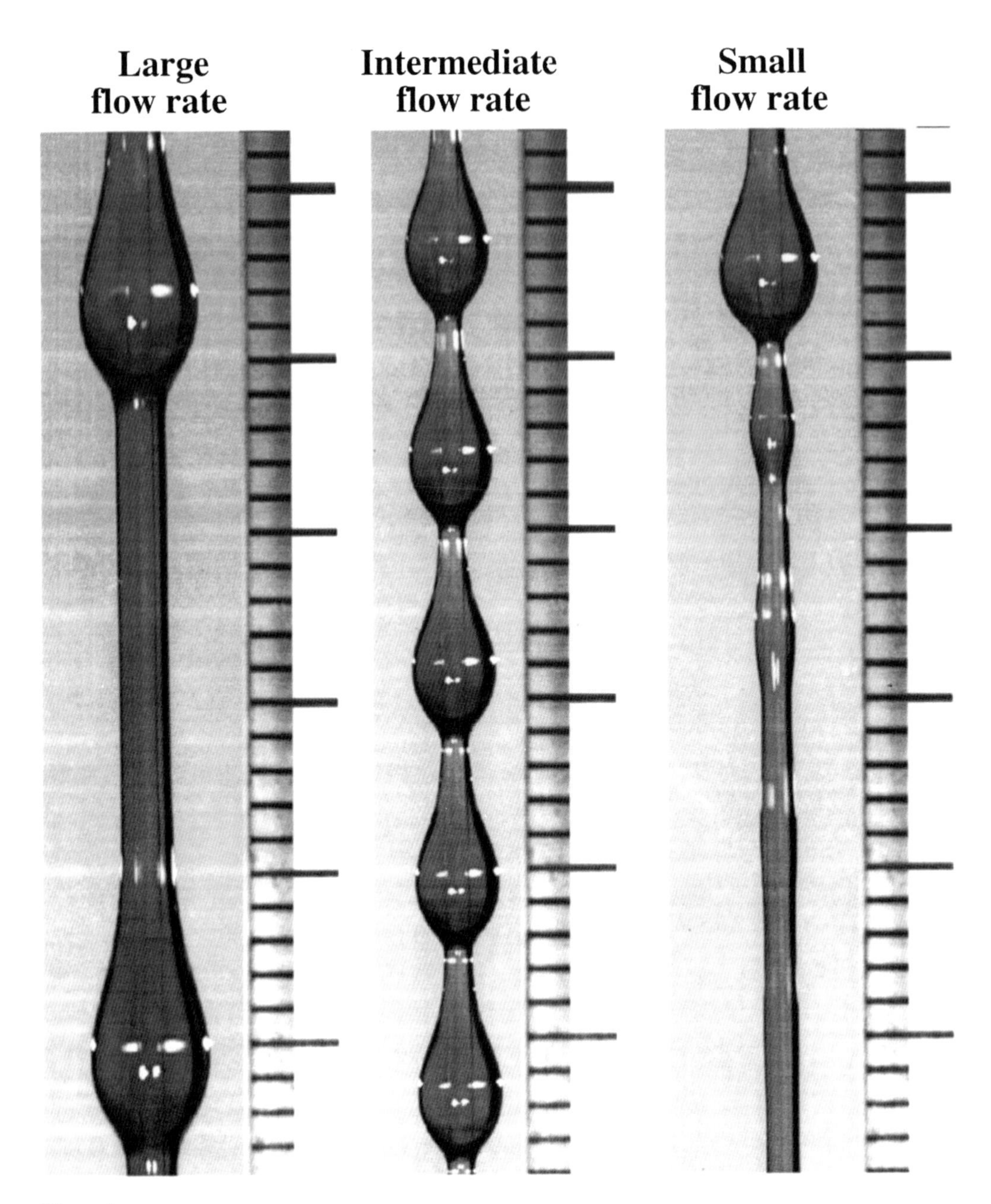

Figure 1

Viscous beads on thin vertical fiber

Igor L. Kliakhandler, Stephen H. Davis and S. George Bankoff

Northwestern University

A thick uniform viscous film of castor oil on a vertical wire is unstable and forms beads. The Reynolds number of the flow is very small, typically 0.01. The small marks on ruler are 1 mm apart. As a fiber, nylon fishing line of radius 0.25 mm was used.

Three distinct regimes of the flow were observed. The first regime is observed for relatively large flow rates. In this case, the drops are large and move rapidly. The film between the drops is relatively thick and practically uniform. In the second case, the drop train is periodic. This regime was observed in a relatively small range of flow rates. For small flow rates, the drops are substantially more separated in space, and the film between the large drops shows a growth of periodic disturbance. Large drops collide with growing lobes ahead of them, consume them, and continue to move.

Keywords
surface tension.

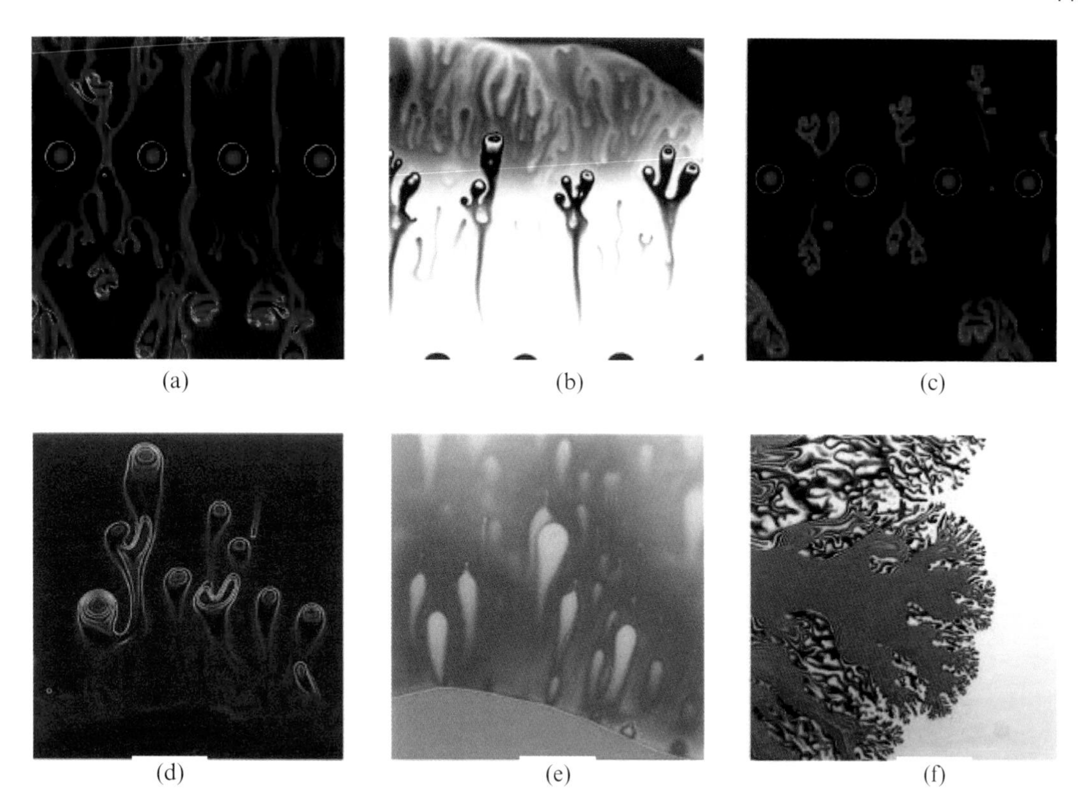

Figure 1

Front propagation and branching dynamics in surfactant driven spreading

Benjamin J. Fischer, Anton A. Darhuber and Sandra M. Troian *Princeton University*

The spreading of a surface active film on a liquid layer of higher surface tension is known to produce fingering instabilities.[1–3] These instabilities resemble dendritic arrays of liquid streamers that undergo repeated branching and tip-splitting. Despite the similarity to viscous fingering patterns,[4] the physical mechanisms governing Marangoni driven spreading are quite different. In particular, the viscosity ratio is found to play little or no role. In these experiments, a 10 μm glycerol layer, with surface tension 63.4 dyn/cm, was first spin-coated onto a cleaned silicon wafer. The wafer was then raised toward a 142 μm diameter chromium-nickel wire that was positioned in the focal plane of an microscope with a 540 nm bandpass filter. The wire was coated with oleic acid of surface tension 32.5 dyn/cm. Upon contact of the two liquids, the oleic acid spontaneously spread on the glycerol layer. The Marangoni shear stress, caused by the gradient in surface tension, rapidly thinned the glycerol, especially near the point of contact. Lifting the wire created a second (Rayleigh-type) instability leading to the linear array of 250 μm diameter droplets shown in Figs. (a)–(c). The oleic droplets act as additional localized sources of surfactant. The six images represent various patterns observed during the spreading process. The field of view is 2.5 mm and the contour lines represent optical interference fringes. Fig. (e) depicts streamers growing just ahead of an oleic droplet in a region thinned by the initial spreading front. Thinner initial liquid layers result in more ramified structures as shown in Fig. (f).

Recent theoretical work on this system has focused on the non-normal (and non-autonomous) character of the linearized operators governing disturbance propagation. These transient growth studies suggest that the Marangoni driven spreading of thin liquid films (*i.e.* within the lubrication approximation) is highly susceptible to perturbations in film thickness and surfactant concentration.[5–7]

Financial support from the National Science Foundation and Unilever Research US is gratefully acknowledged.

Keywords
viscous fingering; spreading; Marangoni stresses.

[1]A. Marmur and M. D. Lelah, *Chem. Eng. Comm.* **13**, 133 (1981).
[2]S. M. Troian, X. L. Wu, and S. A. Safran, *Phys. Rev. Lett.* **62**, 1496 (1989).
[3]M. Cachile and A. M. Cazabat, *Langmuir* **15**, 1515 (1999).
[4]G. M. Homsy, *Annu. Rev. Fluid Mech.* **19**, 271 (1987).
[5]O. K. Matar and S. M. Troian, *Phys. Fluids* **11**, 3232 (1999).
[6]B. J. Fischer and S. M. Troian, *Phys. Rev. E* **67**, 016309 (2003).
[7]J. M. Davis, B. J. Fischer, and S. M. Troian, *to appear* in *Interfacial Fluid Dynamics in Physiochemical Phenomena*, Lecture Notes in Physics, ed. R. Narayanan (Springer-Verlag, 2003).

(c)

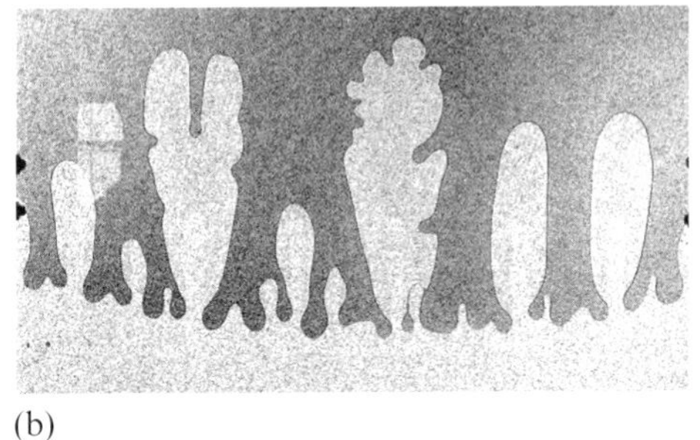

(b)

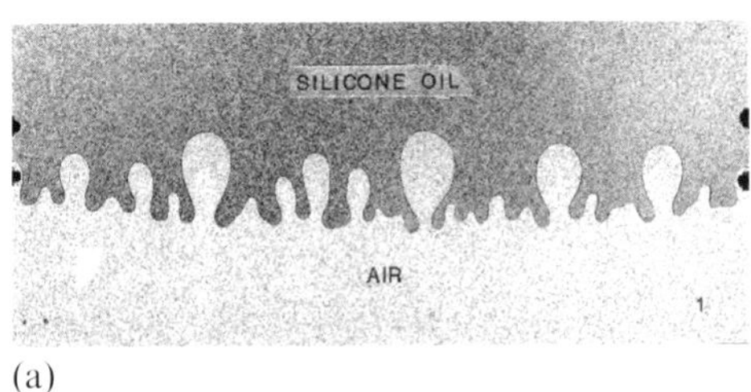

(a)

(d)

Figure 1

A note on the nonlinear growth of a gravitationally unstable interface in a Hele-Shaw cell

T. Maxworthy

University of Southern California

These photographs show the growth of the gravitationally unstable interface between viscous silicone oil ($\nu = 86$ cS) and air contained in the narrow (0.16 cm) gap between two 1.27 cm thick glass plates with working dimensions 61×120 cm. In this case, the cell was rotated until it was vertical, and approximately thirty unstable waves were formed initially. The fastest growing waves suppressed the growth of their smaller neighbors until only one dominant wave emerged. The growth of the effective interface width scales with the wavelength and the growth rate of the initial instability. Casual observation of a highly contorted interface, such as Fig. (d), which is the result of a number of bifurcations, suggests that the interface may have a fractal dimension. By measuring the number of steps (N) required to cover the interface with a gauge of length (G) [Fig. (c)] and plotting log N vs log G, we determine[1] that the interface is one dimensional for scales less than the instability wavelength (L), but has a fractal dimension of 1.36 for scales between L and $10L$. We suspect that running these experiments in a much larger cell would extend this result to even larger gauge lengths. A manuscript outlining the details of a large number of similar experiments is in preparation.[2]

Keywords

Hele–Shaw cell; gravitational fingering; fractal dimension.

[1] B. Mandelbrodt, *The Fractal Geometry of Nature* (Freeman, New York, 1983).

[2] T. Maxworthy, "The Non-Linear Growth of a Gravitationally Unstable Interface in a Hele-Shaw Cell", *J. Fluid Mech.* **177**, 207–232 (1987).

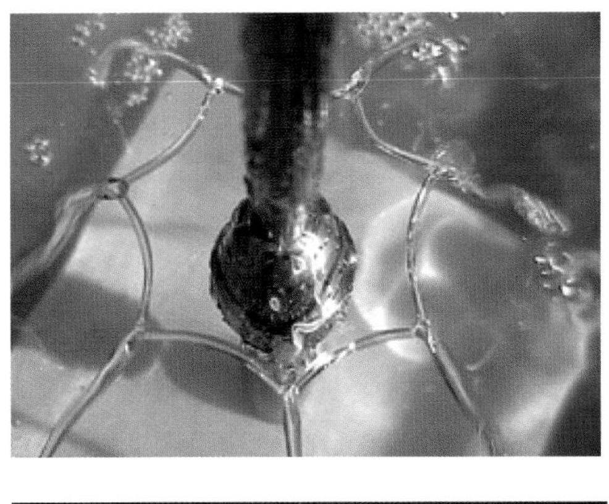
(i)

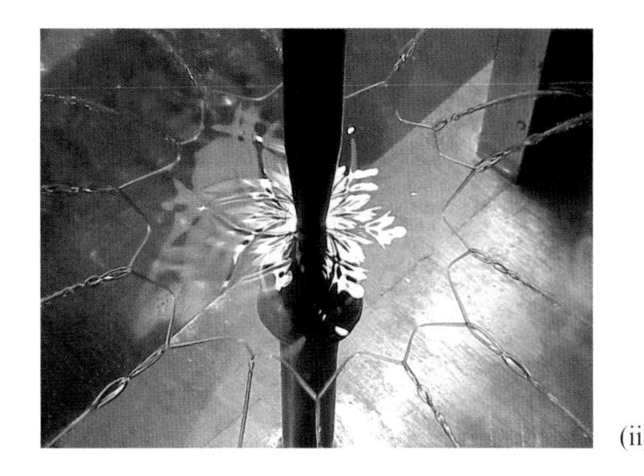
(ii)

(iii)

(iv)

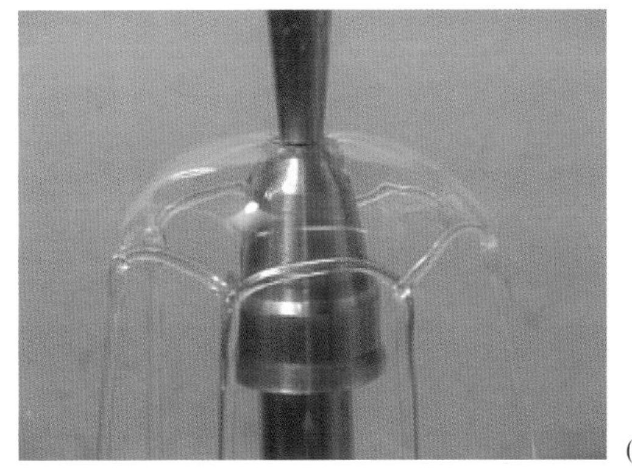
(v)

(vi)

Figure 1

Fluid polygons

Robbie Buckingham and John W. M. Bush
Massachusetts Institute of Technology

We examine the topology of fluid sheets and closed bells generated by extruding viscous fluid radially from an annular gap. The source gap width and radius are typically 1 mm and 3 mm, respectively. Glycerol–water or polyglycol solutions with viscosities between 1 and 100 cS are extruded at typical flow rates of 20 cc/s.

The sheets assume a polygonal form with the number of sides varying from 4 to 16. A heptagonal fluid sheet is illustrated in (i). Fluid proceeds from the sheet to the rim, then streams from the corners of the polygon. Low viscosity sheets ($\nu<10$ cS) support a field of antisymmetrical capillary waves[1] which extend to, and prescribe the form of, the sheet rim (ii). The capillary waves transform the axisymmetric closed bell (iii) into a polyhedral form (iv). The capillary waves are not dynamically significant for the high viscosity sheets ($\nu>50$ cS), where the number of sides is determined by the capillary instability of the toroidal rim. Gravitational deflection of the sheets from the horizontal gives rise to fluid umbrellas (v) and parasols (vi). In certain parameter regimes, the fluid exits the corners of the polygons in the form of a fluid chain[2] (ii).

[1] G. I. Taylor, *Proc. R. Soc. London*, **A 253**, 351 (1959).

[2] M. F. G. Johnson, M. J. Miksis, R. A. Schluter, and S. G. Bankoff, *Phys. Fluids* **8**, S2 (1996).

Keywords
water bells; inertia; surface tension.

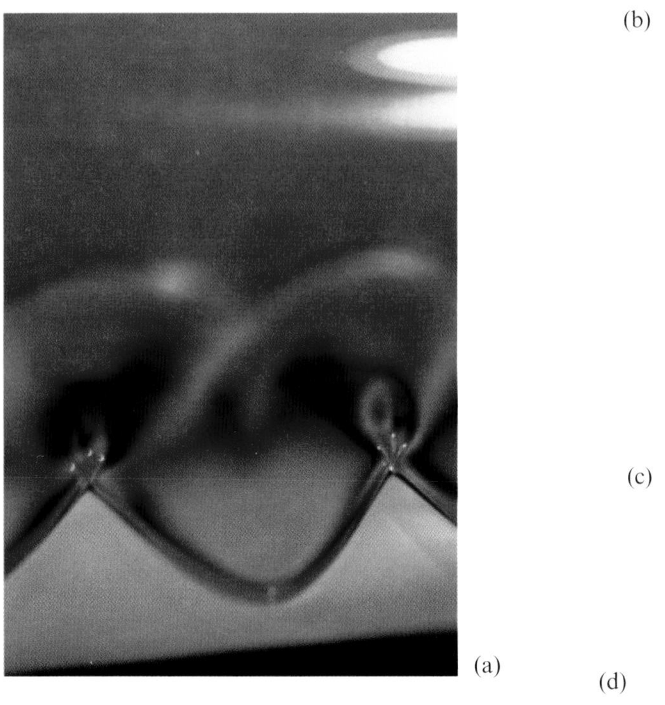

(a)

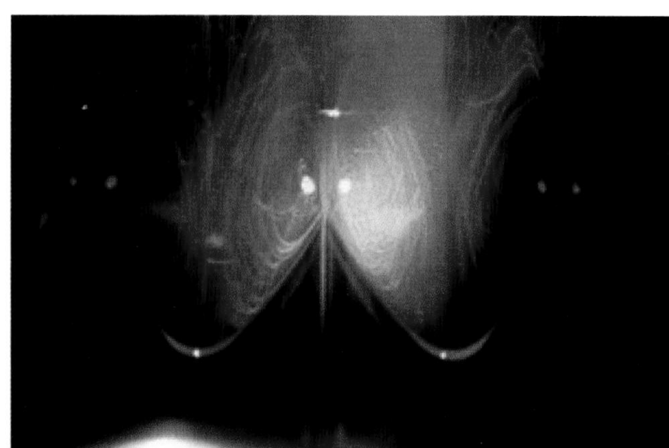

(b)

(c)

(d)

Figure 1

Shark-teeth pattern in coating flow inside a horizontally rotating cylinder

S. T. Thoroddsen and L. Mahadevan
University of Illinois at Urbana–Champaign

A small amount of fluid sits inside a circular cylinder which is rotated about its horizontal axis of symmetry. The interaction between viscosity, gravity, inertia, and surface tension gives rise to a number of interesting patterns.[1] The photographs show various aspects of a striking free-surface pattern which we call "shark teeth," in a cylinder 12.5 cm in diameter filled to a volume fraction of 5.5% with a fluid of viscosity about 50 cP and rotating at a rate of 2.5 c/s. (a) Stationary large-amplitude spanwise undulations with cusps form on the front of the recirculation region at the bottom of the cylinder. (b) These patterns are characterized by a pair of vortices connected to the surface on each side of a cusp. Illumination of the fluid seeded with small particles shows the pathlines surrounding these vortices. The two bright spots correspond to stagnation points on the free surface. (c) A perspective view of the pattern along the length of the cylinder. (d) An isolated cusp on an otherwise straight front, for a slightly different set of parameters, demonstrates that the end effects are not crucial.

Keywords
fingers; immiscible displacement.

[1] S. T. Thoroddsen and L. Mahadevan, "Experimental study of coating flows in a partially-filled horizontally rotating cylinder," *Experiments in Fluids* **23**, 1–13 (1997).

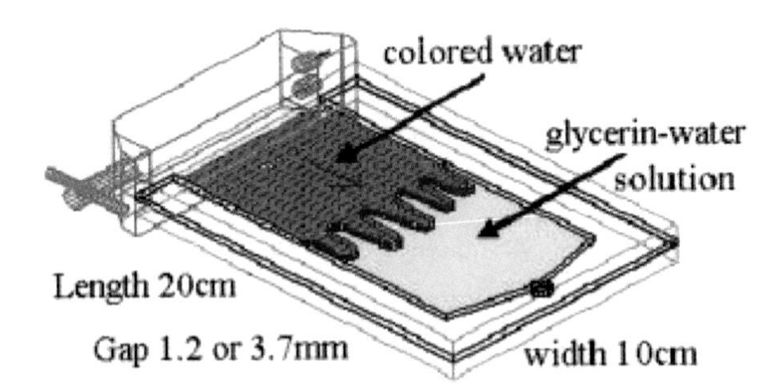

Figure 1 Hele-Shaw cell and the cavity dimensions.

Figure 2 Parabolic manoeuver of the airbus A300 Zero-G.

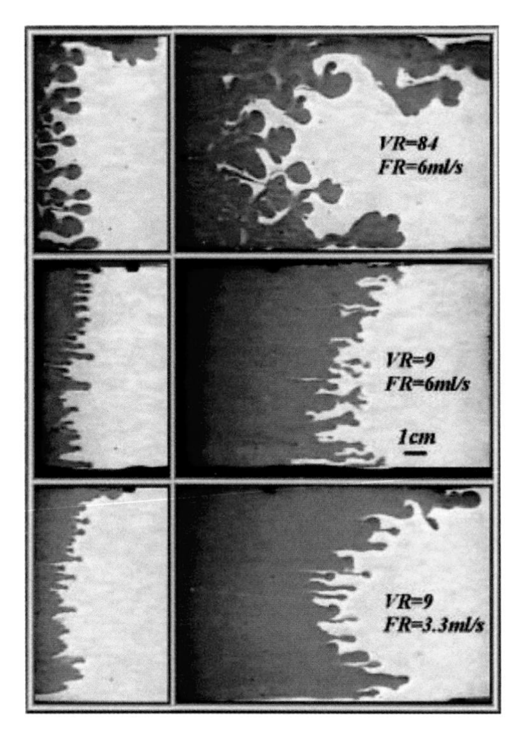

Figure 3 Fingering in microgravity. VR = viscosity ratio. FR = flow rate.

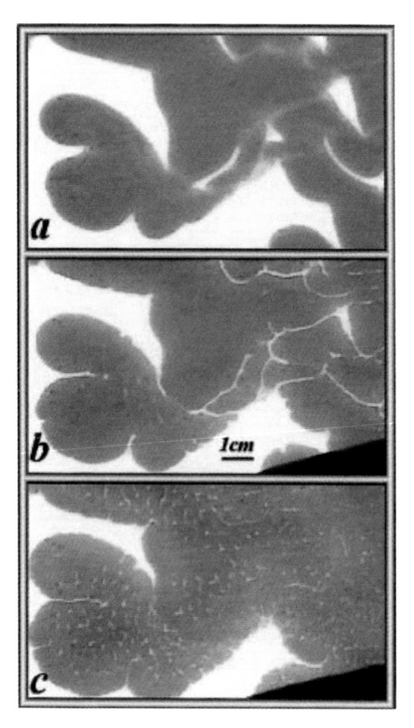

Figure 4 Interpenetration of layers during gravity change from (a) ±0.01 to (c) +1.8 g. No injection.

Viscous fingering in miscible liquids under microgravity conditions

A. A. Vedernikov, B. Scheid, E. Istasse, and J. C. Legros

Université Libre de Bruxelles, Belgium

Viscous fingering was observed by injecting colored water into a Hele-Shaw cell (Fig. 1) preliminarily filled with glycerin–water solution. We varied the viscosity ratio, the flow rate, the gap width, and the density ratio. The Péclet number was higher than 10^5. Part of the experiments was performed in microgravity conditions (parabolic flights) in order to eliminate the gravity influence on pattern formation. Fig. 3 shows peculiarities of fingering in microgravity conditions for a gap width of 1.2 mm. The initial stages are in the left and the developed patterns are in the right column being separated by 1.2 s for the first two pair of images and by 2.4 s for the last one.

Behind the front, the liquids formed a three-layer "sandwich" with the injected liquid in the middle of the gap. During and after transition from micro- to normal gravity, such a system has a tendency to form a two-layer system. The typical stages are shown in Fig. 4, where the gravity vector was perpendicular to the image plane. Here we have: (a) initial state; (b) formation of first holes and long channels in the thinned regions of the low viscosity layer; (c) multiple hole formation far from the periphery of the water layer and squeezing of the channels. The time interval between images (a) and (c) in Fig. 4 was 1.5 s, the density ratio of the liquids before mixing was 1.22, the gap width was 3.7 mm. Finally (not shown), the holes also squeezed, the layers uniformly spread over the entire plane of view and gradually mixed within the gap by diffusion.

ESA, NOVESPACE (France) and OSTC (Belgium) are greatly acknowledged.

Keywords

Hele-Shaw cell; viscous fingering; microgravity.

7 Free surface interaction

Figure 1

Laminar jets can splash!

Mario Errico

University of California at San Diego

It has been observed that a liquid jet impinging on a solid surface can produce splashing. High-speed photography has revealed that, with a turbulent jet, splashing is related to the jet surface roughness. To investigate the importance of the jet shape on splashing, perturbations of known frequency or amplitude are imposed on the surface of a smooth laminar jet.

The top picture shows the unperturbed smooth jet as it spreads radially on the solid surface. The varicose deformations imposed on the jet surface alter the flow quite dramatically (center picture). As we further increase the amplitude of the oscillations, splashing starts suddenly. The bottom picture shows the beauty and complexity of splashing.

Keywords

forced jet; atomization; jet impingement.

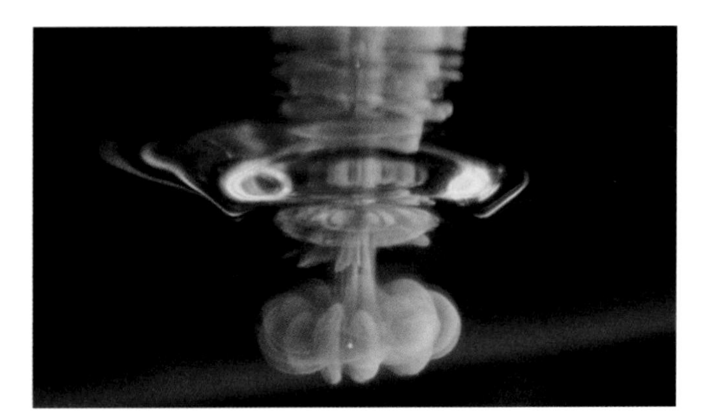

Figure 1 15 msec.

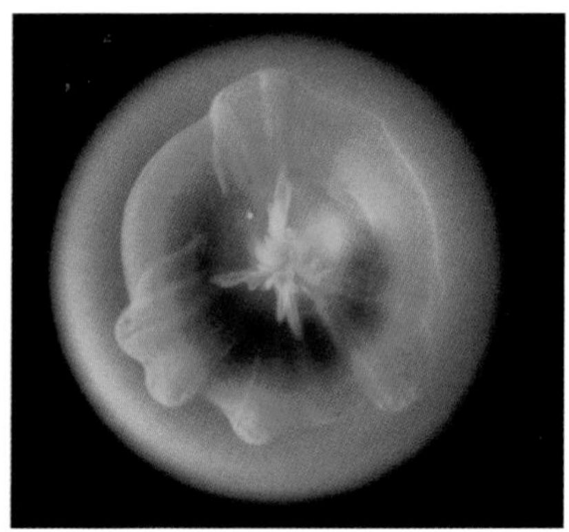

Figure 2 19 msec.

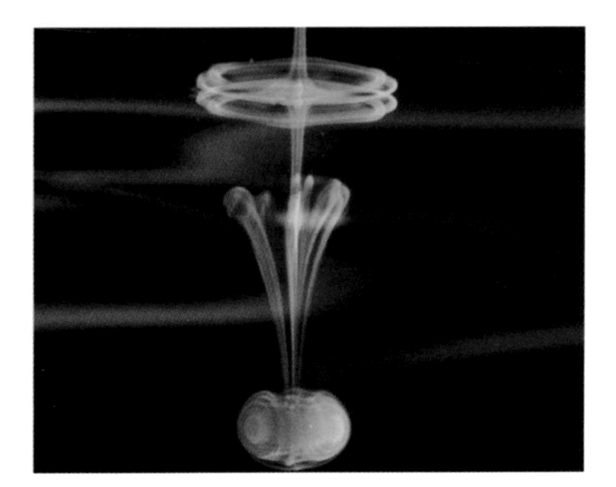

Figure 3 65 msec.

Figure 4 65 msec.

Impacting water drops

Bill Peck and Lorenz Sigurdson

University of Alberta

The four photographs shown here are representative of a series which recorded the structure and evolution of the vorticity generated by a water drop impacting a free surface of water in a container. The 2.8 mm diam water drop was dyed with fluorescein and released from the tip of a hypodermic needle under specific parameters, We = 26, Fr = 25.[1] The Weber number ($\rho U^2 d/\gamma$) and Froude number (U^2/gd) are based on drop diameter d, impact velocity U, and surface tension γ.

Figure 1 is photographed from the side and slightly below the free surface while Fig. 2 is shot looking directly up at the free surface via a mirror. These alternative viewing angles provide a valuable tool in visualizing the three-dimensional flow structure. A "primary" vortex ring can be seen convecting away from the free surface. A convoluted secondary structure can be seen wrapped around the primary core. Vorticity of opposite sign to the primary ring has been produced in the reversing impact crater and has rolled up into concentric rings on the free surface. The sign of the base rings keeps them at the free surface where they interact with their images and increase in diameter.

Figures 3 and 4 show that the vortex structure at later times has the characteristic shape of the atomic blast as discussed by Sigurdson.[2,3] The primary ring has relaminarized having shed the secondary vorticity into the wake to form hairpins or "petals." The azimuthal positions of the petals coincide with the vertices of the perturbed base ring (Fig. 4) suggesting a phase lock within the structure.

The authors thank Mr. Bernie Faulkner for his technical assistance and the Natural Sciences and Engineering Research Council of Canada for their support (Grant No. OGP41747).

[1] B. Peck and L. Sigurdson, "Gallery of fluid motion: Impacting water drop," ed. H. Reed, *Phys. Fluids* **A 3**, 2032 (1991). See also B. Peck and L. Sigurdson, "The three-dimensional vortex structure of an impacting water drop," *Phys. Fluids*, **6**(2), 564–576 (1994).

[2] L. Sigurdson, "Atom bomb/water drop," page 78 of this book.

[3] L. Sigurdson, *Bull. Am. Phys. Soc.* **32**, 2095 (1987). See also L. W. Sigurdson, "Flow Visualization in turbulent large-scale structure research," *Flow Vis. Soc., Japan, Atlas of Visualization*, 3, 99–113 (1997).

Keywords

fluorescent dye; hairpin vortex; mushroom vortex.

Figure 1 Development of disturbance on the jet surface ($t = 18$, 24, 30, 39, and 41 ms).

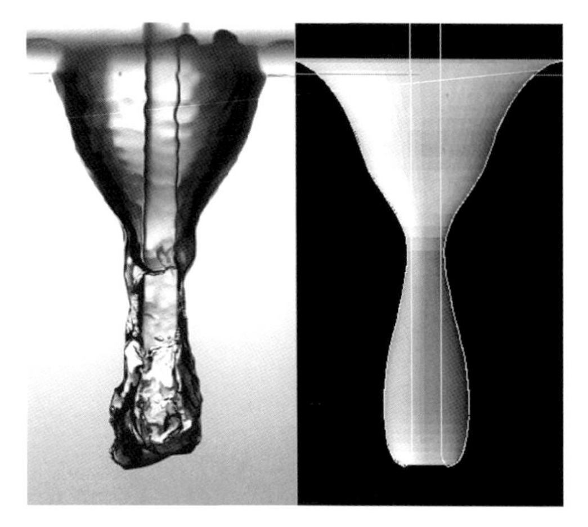

Figure 2 Comparison with numerical simulation.

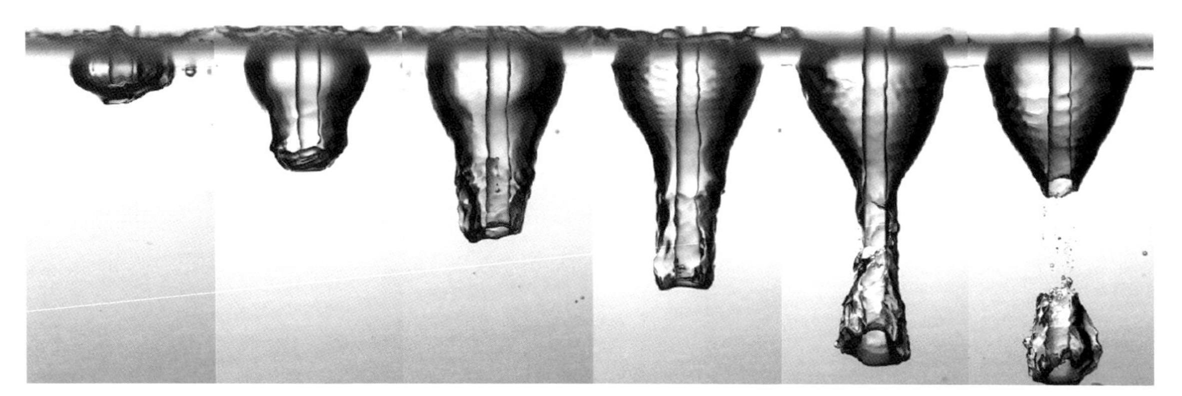

Figure 3 Evolution of underwater air cavity ($t = 47$, 57, 65, 71, 80, and 82 ms).

Air cavity due to jet disturbance

Y. Zhu, H. N. Oğuz, and A. Prosperetti

The Johns Hopkins University

These photographs show the process by which a disturbance on a continuous water jet falling into a water pool triggers air entrainment below the pool surface. The nozzle diameter D is 5.4 mm and the jet velocity U is 1.65 m/s, corresponding to a Reynolds number of 12 300 and a Froude number U^2/gD of 52. By opening the four solenoid valves located upstream of the nozzle, the jet flow rate is rapidly increased by 27%. The increased flow causes the formation of a bulge that develops under the action of gravity and surface tension (Fig. 1). Prior to the arrival of the disturbance, one notices a clear reflection of the jet in the surface and an upward curved meniscus around the jet, both suggesting that no air is entrapped although the Reynolds and Froude numbers exceed threshold values reported by earlier investigators (see, e.g., Bin[1]).

The evolution of the process after the bulge strikes the pool surface is shown in Fig. 2. The depth and width of the surface depression initially increase as the kinetic energy of the bulge and of the jet is transformed into potential energy. After reaching a maximum size, the motion reverses and the lateral surface of the cavity collapses against the jet entrapping a toroidal bubble below the undisturbed water level. The entrapped air with the severed jet inside continues to move downward before breaking up into smaller bubbles, while the remainder of the depression above the pinch-off point rises back toward the surface. Boundary integral simulations of the process have been carried out and found to be in good agreement with the experimental observations, as demonstrated in the example shown in Fig. 3 where the cavity is just about to pinch off. More details are given in reference 2.

This work was supported by the Office of Naval Research.

[1] A. K. Bin, "Gas entrainment by plunging liquid jets," *Chem. Eng. Sci.* **48** (21), 3585 (1993).

[2] Y. Zhu, H. N. Oğuz, and A. Prosperetti, "On the mechanism of air entrainment by liquid jets at a free surface," *J. Fluid Mech.* **404**, 151–177 (2000).

Keywords

air entrainment; cavity collapse; gravity; surface tension.

Figure 1

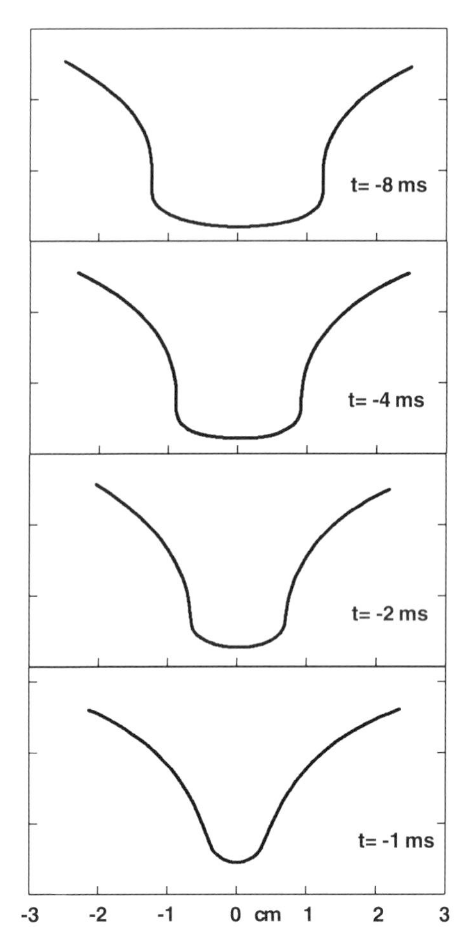

Figure 2

Formation of a self-focusing singularity on a fluid surface

B. W. Zeff and Daniel P. Lathrop

University of Maryland

J. Fineberg

University of Maryland and Hebrew University of Jerusalem

This multiple exposure photograph shows the collapse of a surface wave minimum and the subsequent upward jet due to self-focusing of the kinetic energy associated with a near singularity. Four time steps of the collapse are shown, corresponding to 8 ms, 4 ms, 2 ms, and 1 ms before the singularity; the jet shown corresponds to a time 33 ms after the singularity. The Rayleigh instability causes the tip of the jet to break into droplets and is one source of an ultraviolet cutoff for these near singularities. Digitized surface profiles for the four times preceding the singularity are shown to the right.

Such singularities occur on surfaces undergoing vertical excitation of the form $a = a_0 \sin(\omega t)$ in which the acceleration amplitude a_0 is near a critical value a_c. To create such an event, a glycerin–water mixture with a viscosity of 2.0 cm^2/s is oscillated with a frequency of 7.84 Hz in a periodic wave state ($a_0 < a_c$). At a known time, the acceleration is increased suddenly to a_c, and a singularity occurs 1.4 s later. If the acceleration is increased to a value above a_c, the collapsing minimum entrains an air bubble, and little energy focuses into the jet.

A similarity solution for the collapsing minimum has been found and explains the focusing of kinetic energy and curvature into this structure. Velocities near the tip of the jet can exceed 50 m/s, whereas typical velocities in the pre-jet waves are in the order of 3 cm/s. This work is relevant to understanding breaking surface waves, air bubble entrainment, and jets formed in cavitation bubble collapse and is part of our ongoing study of possible singularity types on a liquid free surface.

We would like to acknowledge support by NSF DMR-9895037.

Keywords

Faraday forcing; Rayleigh instability.

Figure 1

Air entrainment by a plunging jet translating over a free surface

D. Chirichella, R. Gómez Ledesma, K. Kiger, and J. H. Duncan

University of Maryland

The photographs above illustrate the mechanism by which a steady vertical laminar jet ($D_{jet} \approx 6$ mm, velocity $V_{jet} =$ 303 cm/s) induces air entrainment as it translates horizontally over a quiescent pool (velocity $V_t = 44$ cm/s from right to left). In the above surface view (top photograph), the laminar jet appears as a smooth cylindrical rod. Capillary waves are seen propagating upstream of the impact site, while a steep cavity and turbulent free-surface wake are found downstream.

The below-surface view shows the free-surface cavity with air entrainment occurring from the cusp at the bottom of the cavity. The jet entrains no air when it is stationary or translating at speeds below a critical inception value (38 cm/s in the present case). For all observed cases, the inception point followed a Froude scaling given by $Fr = V_t / \sqrt{(g D_{jet})} \approx 1.5$. Visualizations of the flow not shown here indicate that the leading edge of the cavity forms on the downstream surface of the jet. It is believed that the cusp in the near wake of the jet is formed as a result of the low pressure created by the jet's wake combined with the shear induced by the jet's impact velocity. High-speed movies of the entrainment process indicate that the vortex rings of the evolving subsurface jet induce undulations on the leading edge of the cavity. Above the critical conditions for air entrainment, the length of the cavity/jet interface is sufficient to allow the undulations to grow into distended pockets of air. These pockets are subsequently drawn into the vortex ring of the jet and eventually broken up into clusters of bubbles by the turbulent jet.

This work was supported by the National Science Foundation under grant CTS-9876434.

Keywords

air entrainment; bubble formation; capillary wave.

Figure 1 Re = O (HUNDRED)

Figure 2 Re = O (BILLION?) "BIG ENOUGH"

Atom bomb / water drop

Lorenz Sigurdson

University of Alberta

On the right, an aboveground nuclear test in Nevada in 1957 (US Department of Energy).

On the left a water drop falling into a pool of clear water (photo inverted, drop dyed with fluorescein, photo taken by Peck and Sigurdson[1]).

The similarity in large-scale structure is despite a tremendous difference in Reynolds number and buoyancy effects.

The hypothesized vortex skeleton structure[1,2] is represented by three closed vortex lines: the primary ring with four azimuthal waves, four connected loops "shedding" from the ring, and four counter-rotating vortex pairs forming a "stalk" reaching from the primary ring to another ring of opposite sign situated at the bounding surface.

The reasons for the similarity in structure involve the initial conditions for the vorticity generation. The bomb shortly after detonation consists of a fireball and vorticity is generated at its surface. It is produced from pressure gradients acting on the density gradient between the less dense fireball and the nearby air. The pressure gradients consist of the vertical hydrostatic pressure gradient and the pressure gradient associated with the shock wave which is reflected upward from the ground.

The vorticity generation mechanism for the water drop has been an area of investigation.[1,3]

[1] B. Peck and L. Sigurdson, "The three-dimensional vortex structure of an impacting water drop," *Phys. Fluids* **6** (2), 564–576 (1994); L. Sigurdson, "Flow visualization in turbulent large-scale structure research," *Flow Vis. Soc., Japan, Atlas of Visualization*, **3**, 99–113 (1997).

[2] L. Sigurdson, *Bull. Am. Phys. Soc.* **32**, 2095 (1987).

[3] L. Sigurdson and B. Peck, *Bull. Am. Phys. Soc.* **34**, 2286 (1989).

Keywords

mushroom vortex.

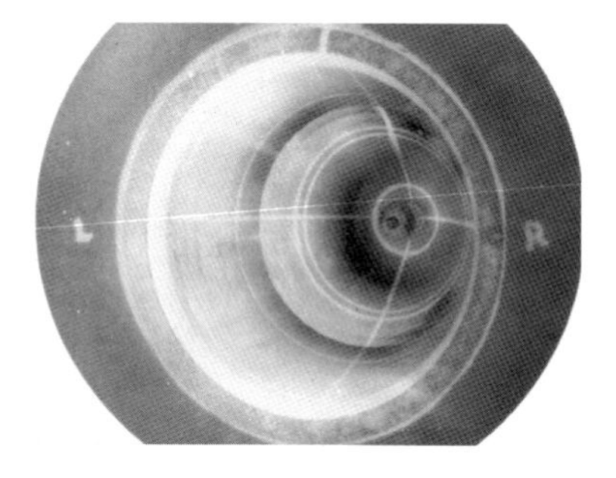

Figure 1 17.1 μs.

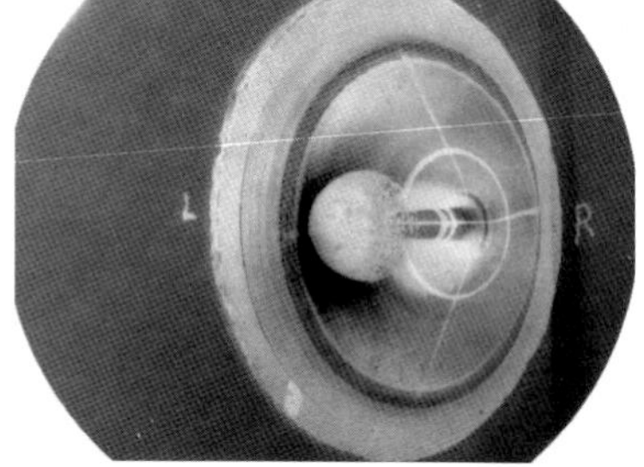

Figure 2 24.1 μs.

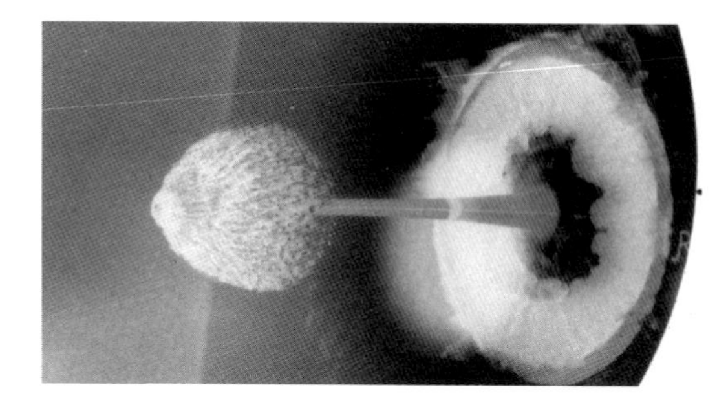

Figure 3 35.8 μs.

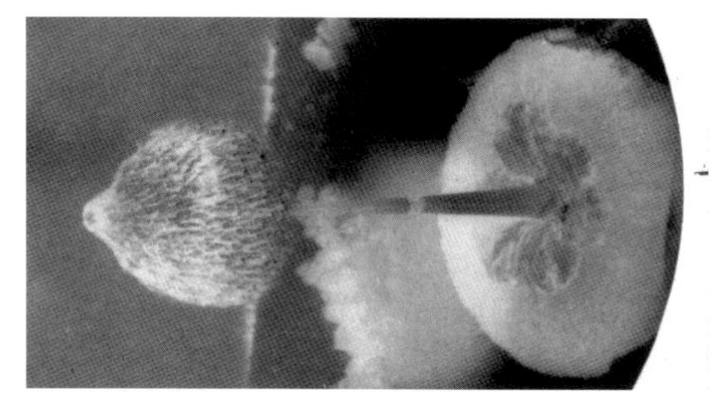

Figure 4 40.8 μs.

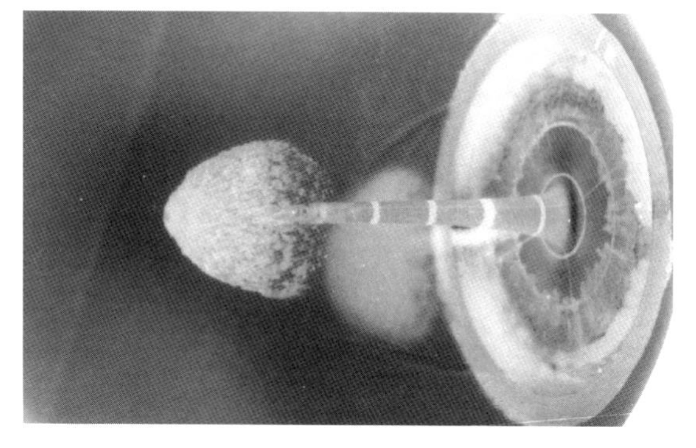

Figure 5 31.5 μs.

Hypervelocity explosive-driven metal jet in air

L. L. Shaw, S. A. Muelder, D. W. Baum, and K.A. Winer

Lawrence Livermore National Laboratory

The above pictures, chronicling the development of a hypervelocity, explosive-driven, copper jet in air, represent the first high resolution images ever made of the detailed structure of such a flow.[1,2] In this experiment, a conical copper liner with a base diameter of 66 mm and apex height of 72 mm, is encased in approximately a half-kilogram of explosive. Upon detonation of the charge, the liner collapses and turns inside out forming a jet. The sequential pictures, Figs. 1–4, illustrate the growth of the jet. The jet tip velocity is 9.2 km/s, with a corresponding Mach number of roughly 27.

The pictures are taken with an eight-frame, electro-optic shuttered camera and pulsed ruby laser illumination.[3] The use of monochromatic illumination and narrow bandpass filtering, which excludes extraneous light such as the shock produced luminous sheath that surrounds the jet, and the very short (nominally 20 ns) exposure time, yields high resolution images of extremely fast phenomena.

Several interesting features of the jet (from another identical experiment) are especially apparent in Fig. 5. The lines that appear on the stem of the jet are ink grid lines initially drawn on the inside of the conical liner. In addition, the shock structure is clearly visible, both as a shadowgraph on the background surface and as a schlieren effect. One shock is associated with the tip of the jet and a second with the trailing edge of the head. Perhaps the most remarkable aspects are the very fine scale features which may be distinguished, including the instabilities near the tip, the breakup of the material of the head, and the small ripples in the stem.

This work was performed under the auspices of the U.S. Department of Energy by LLNL under Contract No. W-7405-Eng-48.

Keywords

liquid-metal jet; jet breakup; shock structure.

[1] D. W. Baum, L. L. Shaw, and C. A. Honodel, "Application of high-resolution optical technique to shaped-charge jet formation," *Proceedings Ballistics '92*, 13th International Symposium, 2, 575–583 (1992).

[2] D. W. Baum, L. L. Shaw, S. C. Simonson, and K. A. Winer, "Liner collapse and early jet formation in a shaped charge," *Proceedings, Ballistics '93*, 14th International Symposium, 2, 13–22 (1993).

[3] L. L. Shaw, S. A. Muelder, A. T. Rivera, J. L. Dunmire, and R. D. Breithauft, "Electro-optic frame photography with pulsed ruby illumination, *Proceedings, 20th International Congress on High Speed Photography and Photonics* (SPIE, Bellingham, WA, 1992), SPIE 1801, pp. 92–105.

Figure 1

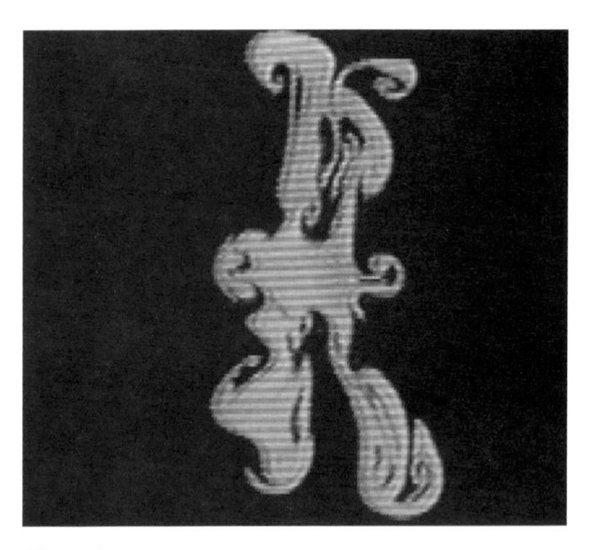

Figure 2

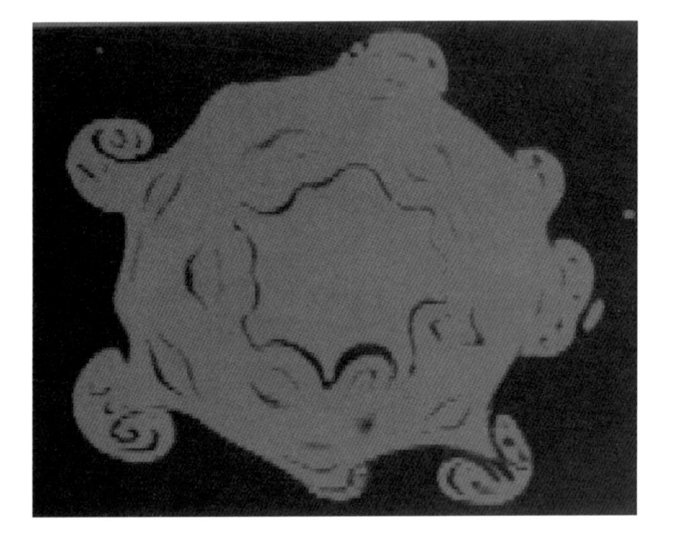

Figure 3

Figure 4

The effect of the free surface on a round jet

D. Liepmann and M. Gharib

University of California, San Diego

The images shown above are sequential cross sections of a round jet at 3.5 diameters downstream of the nozzle visualized using laser-induced fluorescence. Figures 1 and 2 show the flow field of the jet 1 diameter below the free surface, while Figs. 3 and 4 are images of the submerged jet. Except for the presence of the free surface, the flow conditions are identical with a Reynolds number of 4000 based on the jet diameter and exit velocity.

Figures 1 and 3 are vortex rings followed by two images from the braid region before the next primary structures move through the laser sheet (Figs. 2 and 4). In the submerged jet, the azimuthal instabilities grow evenly around the jet core. The secondary structures appear to grow in the braid region between the vortical ring structures that are the primary instabilities of the jet. The streamwise vortex pair that surround the vortex ring in Fig. 1 are attached to the downstream braid/ring structure but have moved out into the slower moving fluid and been left behind their original structure.

The free surface changes the dynamics of both the primary and secondary structures by reducing entrainment from above the flow and creating a stagnation point in the inward flow at the surface above the jet. The jet develops persistent vortex pairs at the top and bottom of the primary structures.

This work has been supported by the Office of Naval Research and DARPA.

Keywords

vortex ring; vortex braids; fluorescent dye.

8 Combustion

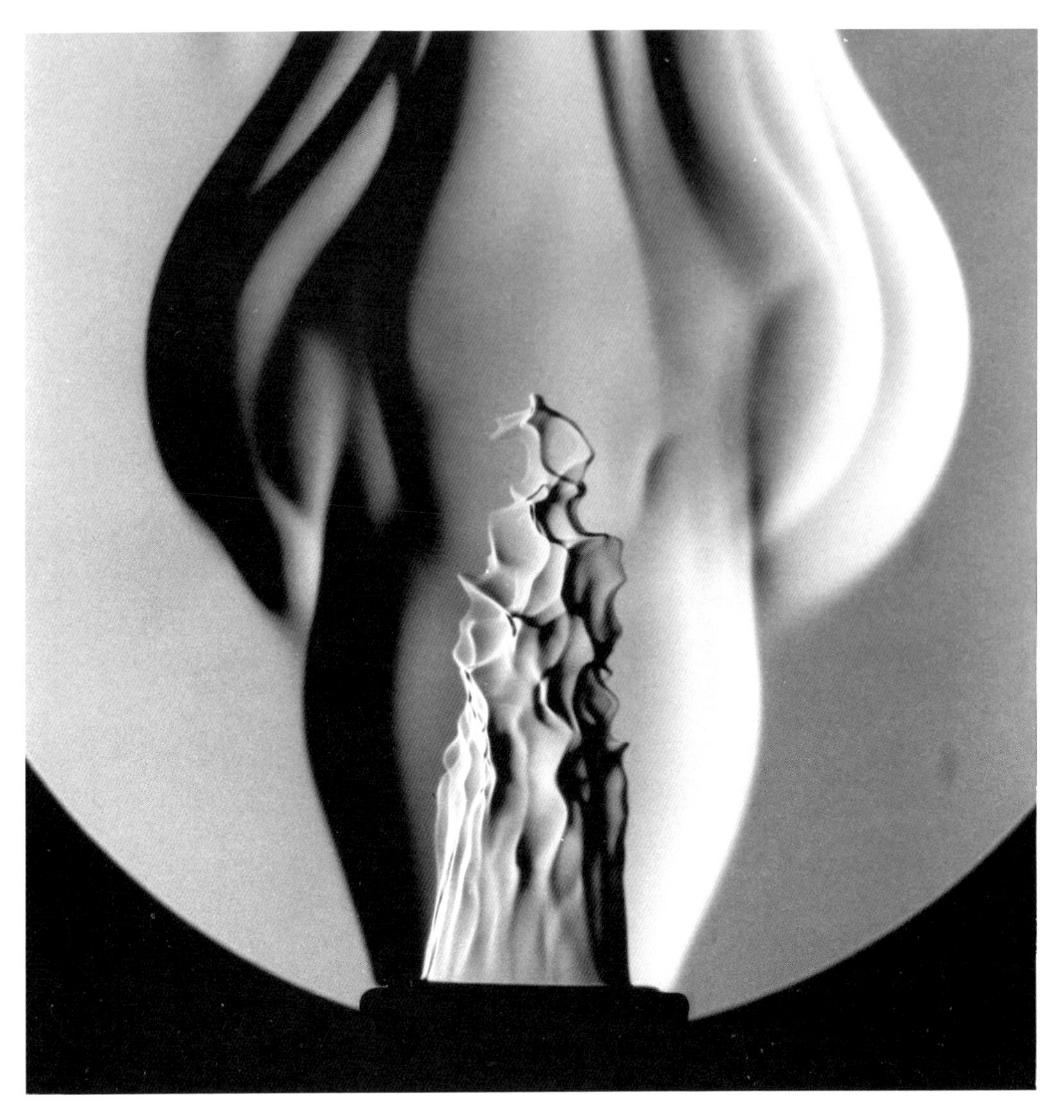

Figure 1

Wrinkled laminar flames

Akira Yoshida

Tokyo Denki University

Burner setup. A propane–air turbulent premixed flame is stabilized on a 30 mm Bunsen-type burner by an annular pilot. The equivalence ratio is 0.68. The flame height is about 85 mm. The mean velocity of the unburned mixture is 2.36 m/sec. Turbulence is given to the mixture by a perforated plate. The turbulence rms fluctuations at the burner exit is 0.15 m/sec. The Taylor and the Kolmogorov microscales are 1.81 and 0.22 mm, respectively, and the Reynolds number based on the Taylor microscale is 17.4.

Photographic setup. This schlieren photograph was taken by a Canon-F1 camera with a 300 mm telephoto lens of $f = 5.6$. Two 200 mm schlieren mirrors with the focal length of 2000 mm were used for the Z-light path arrangement. The vertical knife edge is mounted at the focal point of one mirror. The light source was a xenon stroboscope with a condenser lens and a pinhole. The maximum light power is 8 J (170 1x-sec). The flash duration time is typically 15 μsec. The flash timing was synchronized with the camera shutter. The film used was the Neopan SS (ASA 100) and was developed by Fujidol.

Interpretation. With a moderate or weak turbulence of the unburned mixture, the instantaneous turbulent premixed flame zone consists of a continuous wrinkled laminar flame front. The wrinkle size seems to be irrelevant to the turbulence scale. Along the unburned mixture flow, the amplitude of wrinkles increases from bottom to top. We think that the hydrodynamic instability plays an important role in the flame wrinkling.

Keywords

schlieren; premixed flame.

Re = 408

Figure 1

Re = 1867

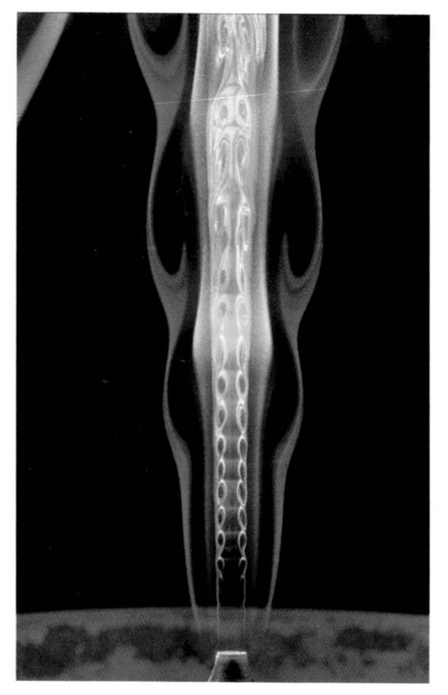

Figure 2

Re = 6420

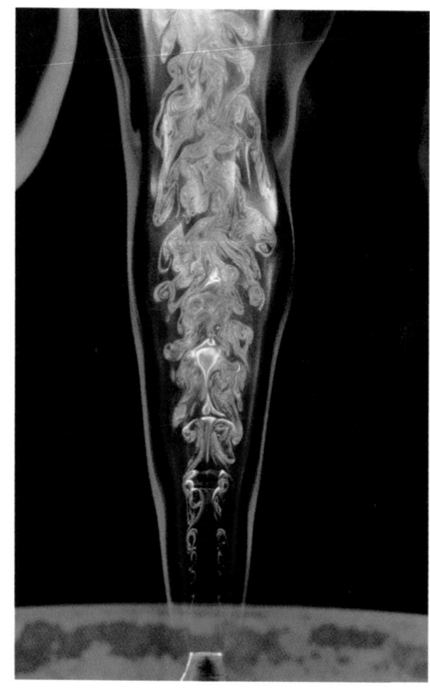

Figure 3

Jet diffusion flame transition to turbulence

W. M. Roquemore

Air Force Research Laboratory

L-D. Chen, J. P. Seaba, and P. S. Tschen

University of Iowa

L. P. Goss and D. D. Trump

Innovative Scientific Solutions, Inc.

A novel laser sheet lighting technique, which can provide very detailed information on the mixing and transport processes occurring inside and outside of flames, is used as a research tool to aid the development of gas turbine combustion design models and in evaluating high-performance–low-pollutant combustor concepts. The technique involves seeding the fuel jet and a coannular jet with $TiCl_4$ vapor. As the fuel and air mix with combustion products, the $TiCl_4$ reacts spontaneously and nearly instantaneously with the water product to form micron-size TiO_2 particles and HCl. Mie scattering from the TiO_2 particles is observed at right angles to the plane of the laser sheet passing vertically through the center of the flame. This provides a view of the flow where the water product mixes with the air outside and the fuel inside the flame. A two-dimensional projection of the flame is viewed in conjunction with the Mie scattered light.

The three methane jet diffusion flames shown here have an annulus air velocity of 0.15 m/s. The methane jet velocity is varied to achieve the cold-flow exit plane Reynolds number shown above each flame. The jet nozzle, having an exit diameter of 10 mm, is contoured to achieve a flat-top velocity profile. The light sheet is formed with a Nd:YAG laser. The camera shutter is triggered by the laser firing. The exposure time is 10 ns for the laser pulse and 2 ms for the yellow flame.

The flames shown here have large toroidal vortices just outside the flame surface. We believe that these buoyancy-driven vortices are responsible for low-frequency instabilities of the flame surface, the 12 Hz flame flicker frequency observed at low Reynolds number, and the wrinkled flame appearance at high Reynolds numbers. The shear layer inside the low Reynolds number flame has a weak instability. At a moderate Reynolds number, this inner instability has developed into coherent vortices. At a high Reynolds number, the coherence of the inner vortices are broken down by coalescence of the vortices. Small, randomly distributed vortices are then observed.

This work was supported by the Air Force Office of Scientific Research, Aerospace Sciences Division.

Keywords

Mie scattering; laminar–turbulent transition; toroidal vortices.

Flame-vortex interactions in a driven diffusion flame

K. Y. Hsu, V. R. Katta, L. P. Goss, and D. D. Trump

Innovative Scientific Solutions, Inc.

L. D. Chen

University of Iowa

W. M. Roquemore

Air Force Research Laboratory

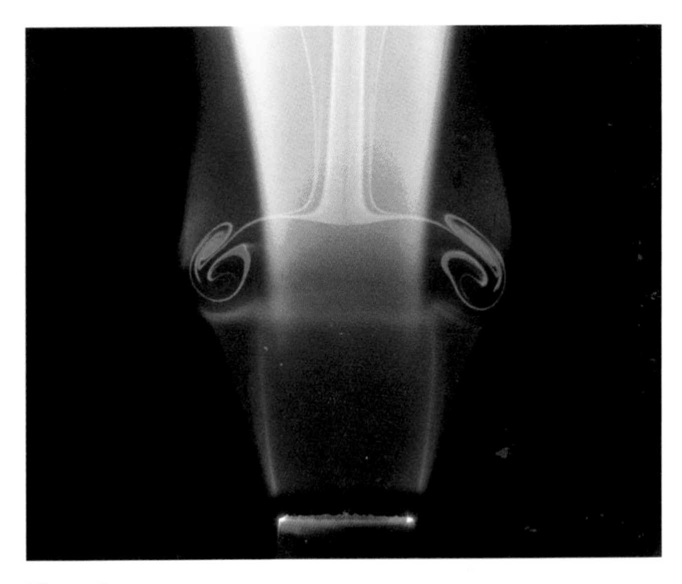

Figure 1

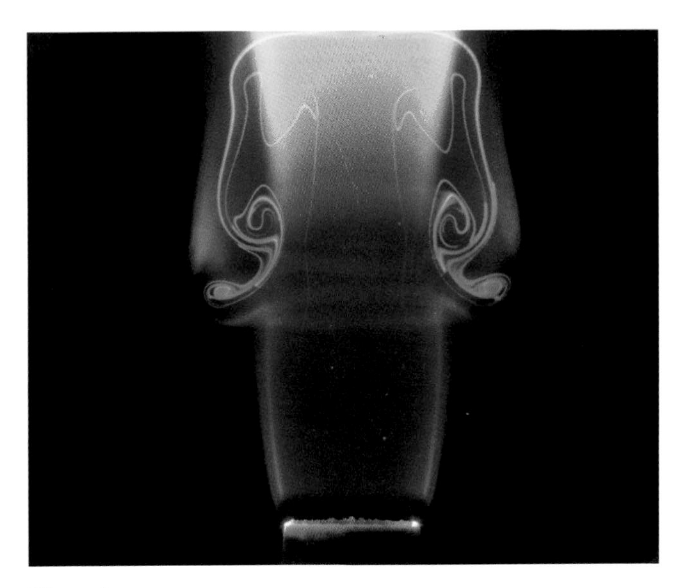

Figure 2

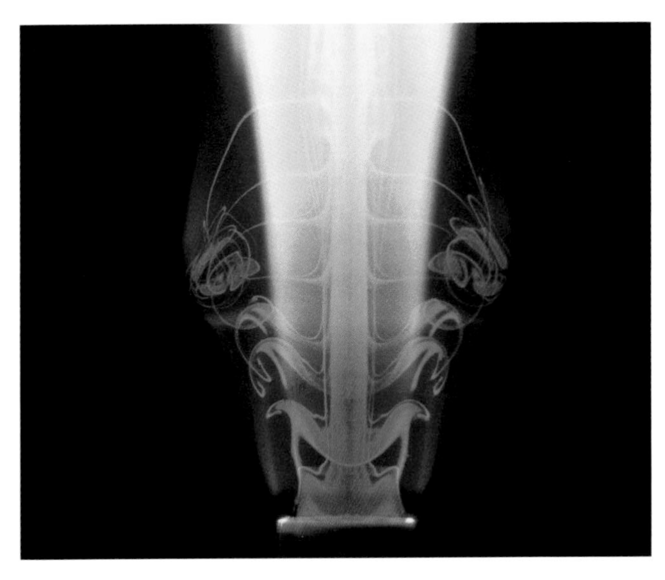

Figure 3

Flame-vortex interactions in a jet diffusion flame are studied in a controlled experiment.[1] The vortices are periodically generated in a methane jet diffusion flame at the frequency of 30 Hz. The phase-locked reactive-Mie scattering technique[2] is used to visualize the interactions between the vortices and the flame. The second harmonic output (532 nm) of a pulsed Nd:YAG laser is expanded to a thin laser sheet vertically across the centerline of the jet. The scattering particles, TiO_2, are formed as a result of the reaction of combustion product H_2O and $TiCl_4$ vapor added to the fuel. The sequential images (Figs. 1 and 2), with 3 msec separation, show the complex dynamics of the interactions between generated vortices and the flame. The large vertical structure containing fuel pushes the flame surface outward in the radial direction. When the flame is stretched, the local flame extinction occurs as is evident by the disappearance of the blue flame. To illustrate the evolution of flame-vortex interactions, an image consisting of eight consecutive phase angles (time interval of 1 msec) is shown in Fig. 3.

This work was supported by the Air Force Office of Scientific Research, Aerospace Sciences Division.

Keywords

Mie scattering; stretched flame; flame extinction.

[1] K. Y. Hsu, L. D. Chen, V. R. Katta, L. P. Goss, and W. M. Roquemore, "Experimental and numerical investigations of the vortex-flame interactions in a driven jet diffusion flame," 31st Aerospace Meeting and Exhibit, *AIAA* Paper No. 93–0455, 1993.

[2] L. D. Chen and W. M. Roquemore, "Visualization of jet flames," *Combust. Flame* **66**, 81 (1986).

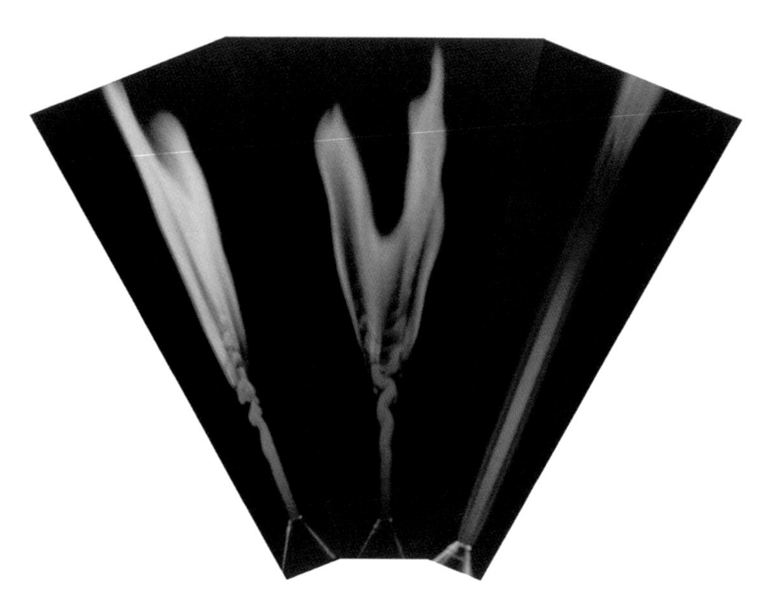

Figure 1

Bouquet with bifurcating jet diffusion flame

Eberhard Pfizenmaier
DLR Institute for Turbulence Research, Germany

Jonathan Simon and Peter A. Monkewitz
University of California, Los Angeles

It has long been known[1] that jet diffusion flames and plumes can be dramatically modified by sound. Flame bifurcations in particular have been documented by Schneider[2] for rectangular fuel jets with acoustic forcing through the nozzle. The first explanation of the bifurcating phenomenon has been given by Lee and Reynolds[3] in terms of vortex induction for the case of a jet excited by *two frequencies* (an axisymmetric fundamental and a "flapping" subharmonic for bifurcation). Here we show that *single-frequency* acoustic forcing can also induce the bifurcation of, in our case, a round laminar jet diffusion flame. The unforced flame is shown on the right frame of the "bouquet," while the center and left frames show two orthogonal views of the forced flame. The fuel is a hydrogen-helium mixture, seeded with titanium dioxide, which exits from a 7.5 mm nozzle into air. Visualization is by a laser sheet through the jet axis and the forcing is effected by a speaker pointed at the nozzle from a radial distance of about 1 m. the experimental conditions are $U_{jet}=26$ m/sec; $\rho_{jet}/\rho_{air}=0.11$ (cold exit conditions); $Re=1.7\times10^3$ (cold exit conditions); mass fraction of hydrogen in the H_2–He fuel jet = 0.22; stoichiometric fuel/air ratio = 1.2 by volume; acoustic forcing at $St=fD/U_{jet}=0.19$ and approximately 110 SPL dB. More details can be found in Monkewitz et al.[4]

From the "corkscrew" appearance of the jet in *both* orthogonal views it is deduced that the instability mode excited by the speaker is a spiral mode. The flame bifurcation in this experiment occurs in a plane parallel to the acoustic wave vector and is therefore associated with the excited spiral mode. It is speculated that the phenomenon is due to a modulation of the spiral vortex strength with the same frequency as the vortex passage frequency (and forcing frequency). This hypothesis, which would again provide an explanation in terms of vortex induction, is currently being investigated.

The support of this study through the ONR-URI on Applications of Complex High Reynolds Number Flows is gratefully acknowledged.

Keywords
hydrogen–helium flame; acoustic forcing; spiral mode instability; Mie scattering.

[1] J. Tyndall, *Sound – A Course of Eight Lectures Delivered at the Royal Institution of Great Britain* (Longmans, Green & Co., London, 1867).

[2] P. E. M. Schneider, "Experimentelle Untersuchungen ueber den Einfluss von Schall auf Diffusionsflammen," *Z. Flugwiss.* **19**, 485 (1971).

[3] M. Lee and W. C. Reynolds, "Bifurcating and blooming jets," *Fifth Symposium on Turbulent Shear Flows* (Springer-Verlag, Berlin, 1985), pp. 1.7–1.12.

[4] P. A. Monkewitz, J. Simon, and E. Pfizenmaier, "On the instability behaviour of laminar hydrogen and hydrogen–helium jet diffusion flames," Report DLR IB 22214–92/B6, 1992.

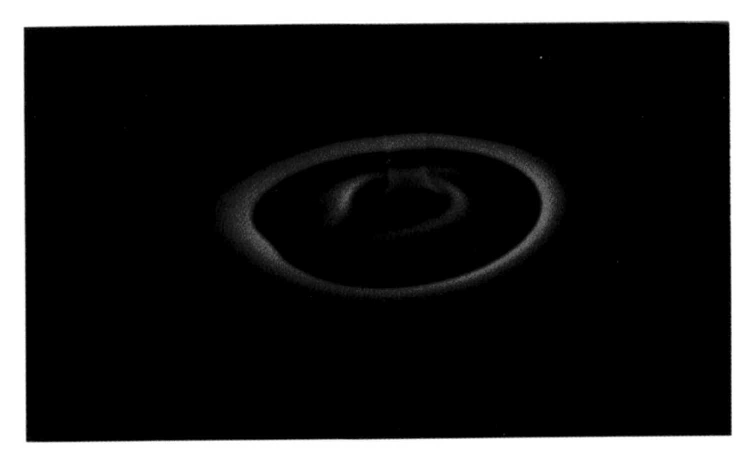

Figure 1(a)

Figure 1(b)

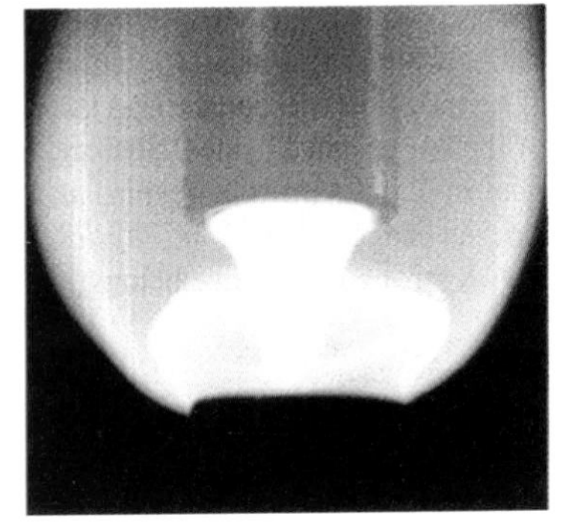

Figure 1(c)

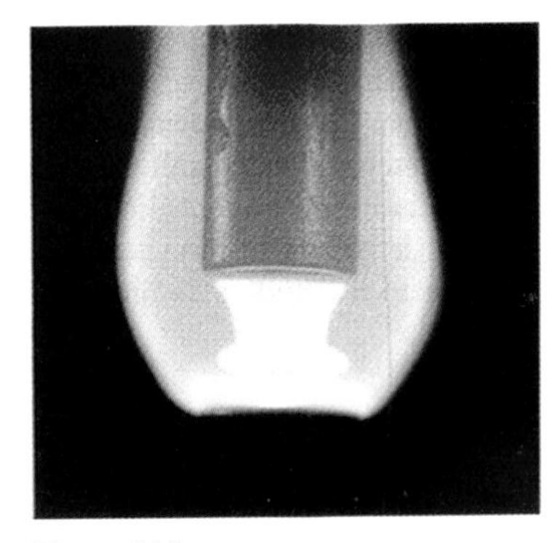

Figure 1(d)

Flame impingements

Nasser Ashgriz

State University of New York at Buffalo

The photographs represent the impingement of methane jet flames on a plate or on each other. Figure 1(a) shows the flame rings generated by the impingement of a single *diffusion* methane flame on a plate from below with Re= 3×10^3. Figure 1(b) shows the impingement of two jets on a plate and the coalescence of the flame rings formed by each jet. The jets are positioned 5 cm from the plate and 4.5 cm from each other. Figures 1(c) and 1(d) show the impingement of two *premixed* methane flames on each other. The Reynolds numbers of the lower jets in Figs. 1(c) and 1(d) are 20 600 and 20 800, and those of the upper jets are 19 300 and 18 200 correspondingly.

Keywords

methane flame; flame impingements.

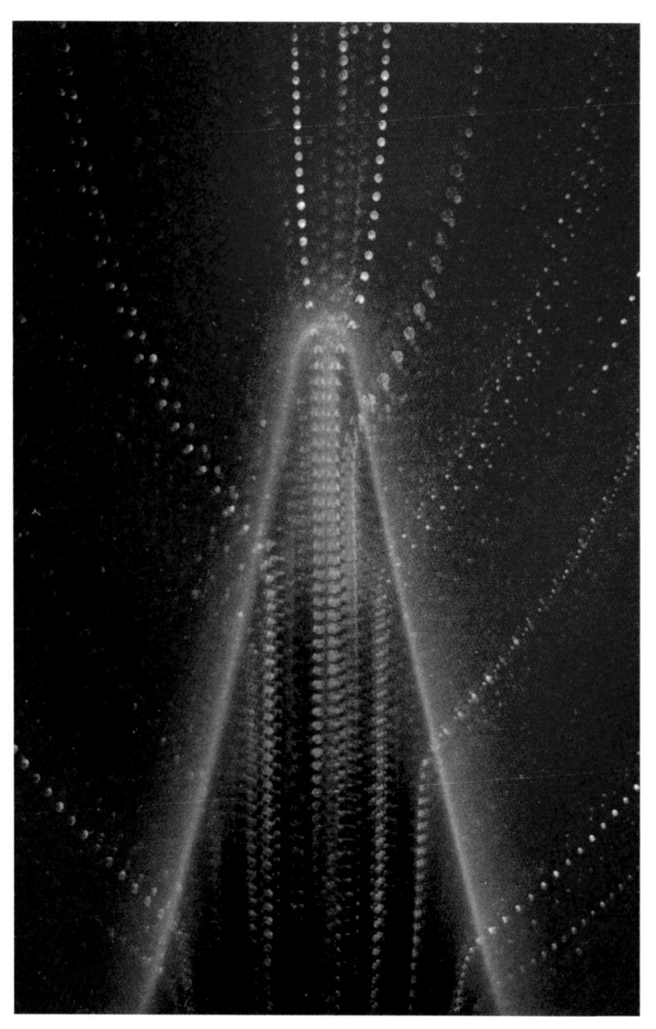

Figure 1

Figure 2

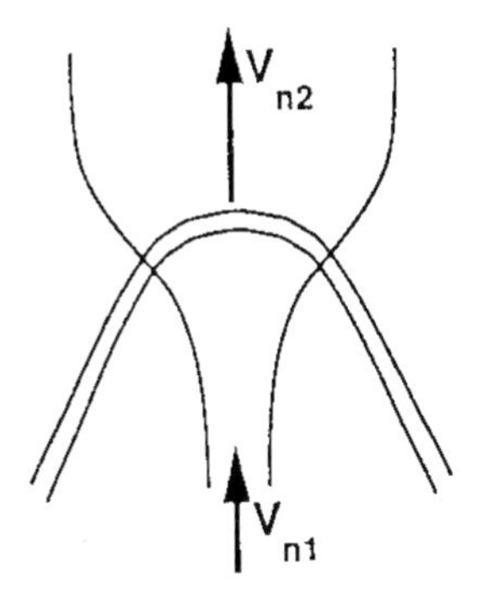

Figure 3

Particle tracking in a laminar premixed flame

T. Echekki and M. G. Mungal

Stanford University

The photograph (Fig. 1) shows a laminar premixed methane–air flame, equivalence ratio of 1.2, stabilized on a two-dimensional slot burner. The flame width at the base is 7 mm. The mean flow velocity is 1.5 m/sec. The flame geometry is visualized by the emission from excited radicals at the reaction zone. The streamlines are visualized by particle tracks using the green light scattered from MgO particles (1–5 μm diameter) carried in the flow. The light from a 20 W copper vapor laser formed into a sheet and operating at a 6 kHz pulse rate is used for illumination. A Nikon camera fitted with an extension bellows and microlens, and Fuji 1600 ASA color film are used to capture the image.

At the side of the flame (Fig. 2), the normal component of velocity (defined as the laminar flame speed) is 0.4 m/sec and the fluid in a streamtube undergoes rapid, constant area acceleration in the normal direction to accommodate the thermal expansion. The tangential component of velocity is unchanged. This is to be contrasted with the streamtube at the flame tip (Fig. 3) where the normal component of the fluid velocity undergoes a small amount of acceleration but there is considerable lateral flow expansion to accommodate the thermal expansion. The flame speed at the tip can exceed the flame speed at the side by factors of 2 to 10 depending upon the amount of lateral expansion.[1]

[1] T. Echekki and M. G. Mungal, "Flame speed measurements at the tip of a slot burner: effects of flame curvature and hydrodynamic stretch," in *Twenty-Third Symposium (International) of the Combustion Institute*, The Combustion Institute, 455–461 (1990).

Keywords

methane flame; streamline; streamtube; Mie scattering.

9 Instability

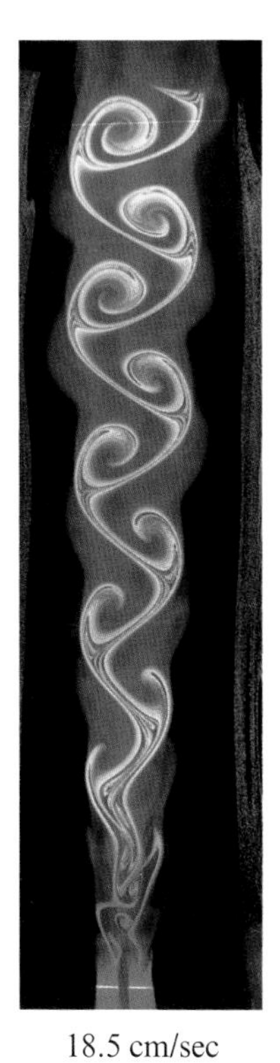

18.5 cm/sec

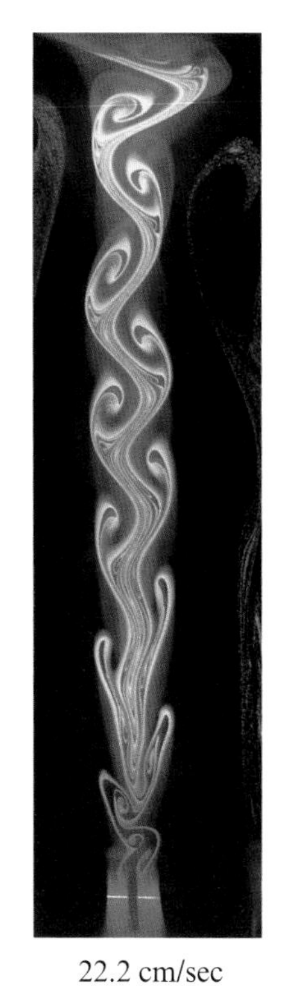

22.2 cm/sec

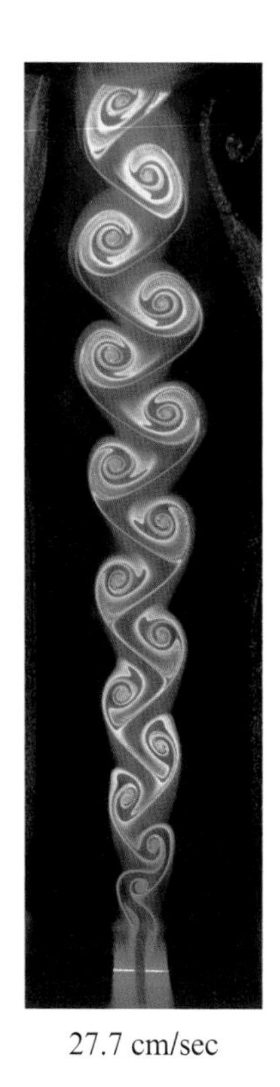

27.7 cm/sec

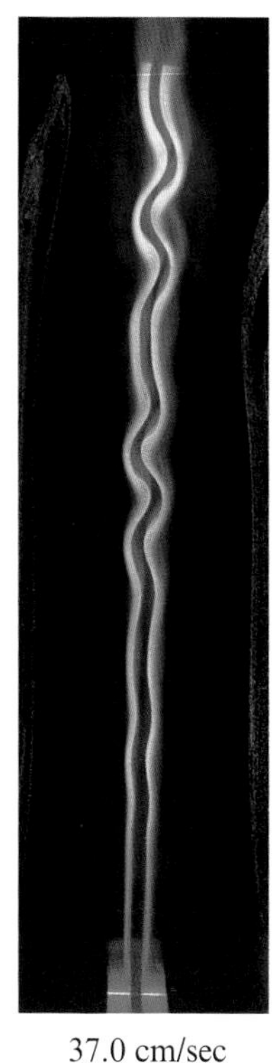

37.0 cm/sec

55.5 cm/sec

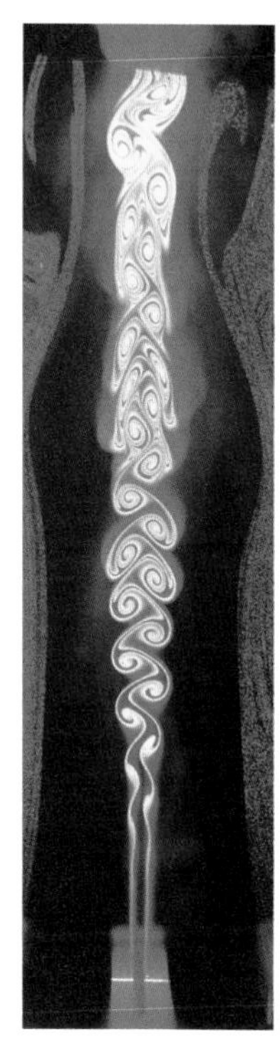

71.0 cm/sec

Figure 1

Interaction of 2D wake and jet plume

W. M. Roquemore and R. L. Britton
Air Force Research Laboratory

Richard S. Tankin *Northwestern University*

C. A. Boedicker *Air Force Institute of Technology*

M. M. Whitaker and D. D. Trump
Innovative Scientific Solutions, Inc.

These photographs show the vortex structures that result from the interaction of vortices that are shed from a 2D bluff body and those shed from a slot jet.[1] The slot jet (3 mm×150 mm) is located in the center of the rectangular face of the bluff body (15 mm×240 mm). The photographs are positioned so that the velocity of the slot jet increases from left to right. In the first three photographs starting from the left, the velocity of the jet is smaller than the velocity of the flow around the bluff-body. In the fourth picture, the shear layer velocities of the jet and bluff body are nearly equal and a wavy structure is observed. At higher velocities, as noted by the 5th and 6th photographs, the vortex structures from the jet dominate the flow field. This is noted by the change in the direction of rotation of the vortices.

The flow is visualized by the Reactive Mie Scattering (RMS) technique in which Mie scattering is observed from micron size TiO_2 particles that are formed by the spontaneous reaction of $TiCl_4$ vapor in the slot jet air with the water in the annulus air. The technique has been shown to be more effective than smoke because it highlights the streamlines where molecular mixing is taking place. The photographs were taken in the 15ns firing of a YAG laser used to form the light sheet.

For an averaged air jet velocity of 18.5 cm/s, the alternating vortex structures shed from the 2D bluff body are evident after about 5 bluff-body widths downstream. As the jet velocity increases, the wake from the bluff body is significantly modified. At a jet velocity of 37 cm/s, the shear layer velocities of the jet and wake are nearly equal and the Kelvin–Helmholtz instability does not grow rapidly. At higher velocities, the jet begins to dominate as noted by the change in the direction of rotation of the vortices.

This work was supported by the Air Force Office of Scientific Research, Aerospace Sciences Division.

[1] C. A. Boedicker, J. P. Planeaux, T.-H. Chen, and J. Schmoll, "Interaction between jet and annulus flows in the presence of a two-dimensional bluff body," presented at First National Fluid Dynamics Congress, Cincinnati, OH, July 1988.

Keywords
Mie scattering; Kelvin–Helmholtz instability; bluff body.

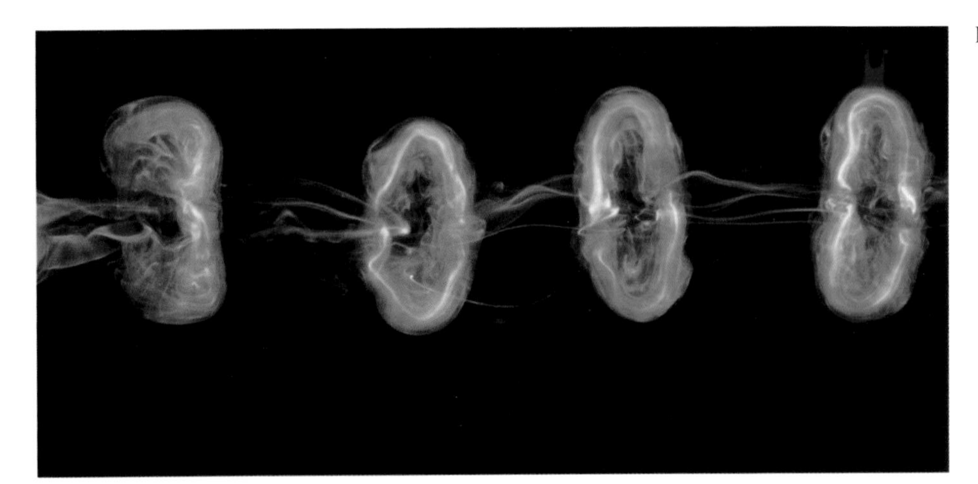

Figure 1

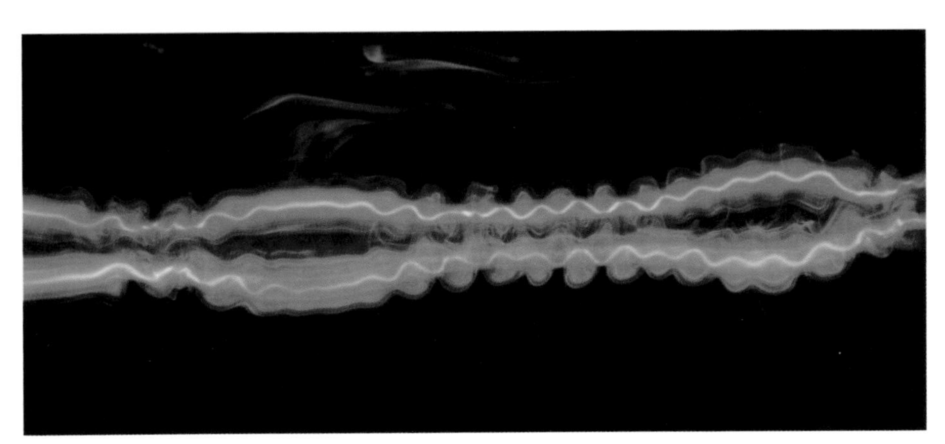

Figure 2

The long and short of vortex pair instability

T. Leweke and C. H. K. Williamson
Cornell University

In these photographs we show visualizations of different instabilities of a pair of straight counter-rotating vortices. Although analytical studies of vortex pair instabilities have suggested the existence of long and short wavelengths,[1,2] it is quite surprising that there exists, in the literature, no clear evidence of such structure in the laboratory.

Here, the vortex pair is generated at the sharpened edges of two flat plates, hinged to a common base and moved in a prescribed symmetric way. Visualization is achieved using fluorescent dye. The evolution of the vortex pair was found to depend strongly on the vortex velocity profiles, which are determined by the motion history of the plates. We identified at least three different length scales from our experiments, two of which can be seen in the photographs.

Long-wave instability. The upper image shows a plan-view of the late state of an instability whose axial wavelength is several times the (initial) distance between vortex centers. The initially straight vortices develop a waviness (similar to the long-wavelength deformations in the lower picture), which is amplified until they touch, break up, and reconnect to form periodic vortex rings, which then elongate in the transverse direction. This late-time evolution has not been observed or studied previously in the laboratory.

Short-wave instability. The lower image shows the development of a short-wave instability (wavelength less than one vortex separation) superimposed on the long waves. The remarkably clear visualization of the vortex core reveals its complicated internal structure, and the observed phase relationships show that the symmetry of the flow with respect to the midplane between the vortices is lost.

This work is supported by the Deutsche Forschungsgemeinschaft (Grant No. Le 972/1–1) and the Office of Naval Research.

[1]S. C. Crow, "Stability theory of a pair of trailing vortices," *AIAA J.* **8**, 2172 (1970).

[2]S. E. Widnall, D. B. Bliss, and C.-Y. Tsai, "The instability of short waves on a vortex ring," *J. Fluid Mech.* **66**, 35 (1974).

Keywords
fluorescent dye; periodic vortex rings; counter-rotating vortices.

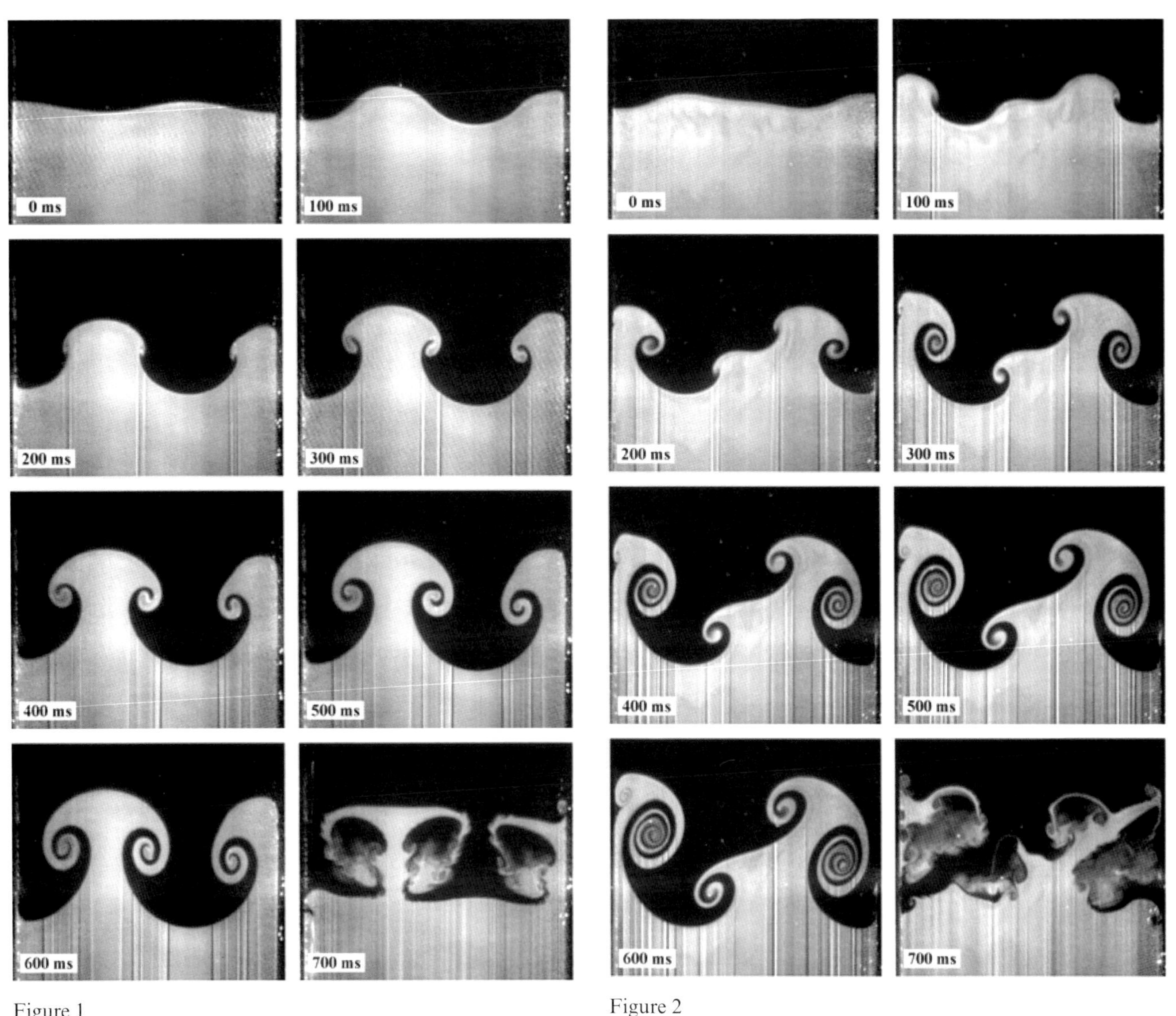

Figure 1

Figure 2

Instability of an impulsively accelerated liquid/liquid interface

C. E. Niederhaus and J. W. Jacobs

University of Arizona

Planar Laser-Induced Fluorescence (PLIF) is used to visualize two different density fluids after they have undergone an impulsive acceleration normal to the interface. The fluids are in a sealed container 120 cm wide which is initially oscillated horizontally to produce a standing internal wave. The container is released from a height of 0.65 m and falls vertically on rails until bouncing off of a fixed spring, imparting an upward impulsive acceleration. The interfacial instability evolves in free fall as the container travels on the rails until hitting the spring a second time. The images are digitized and stored by a computer from a CCD camera traveling with the container. The two sets of images above show the evolution of the instability for different initial conditions. The single-mode initial condition consists of a mode with 1.5 wavelengths within the container (Fig. 1), and the multi-mode initial condition is the superposition of the modes having 1.5 and 2.5 internal wavelengths (Fig. 2). For both sets of images, the first picture shows the initial waveform just prior to impact. The pictures then show the evolution of the instability, and the last picture is taken just after the second impact with the spring. The total time between bounces is 670 ms.

This work is supported by Lawrence Livermore National Laboratory and by NASA Glenn Research Center.

Keywords

Richtmyer–Meshkov instability; Rayleigh–Taylor instability; laser-induced fluorescence.

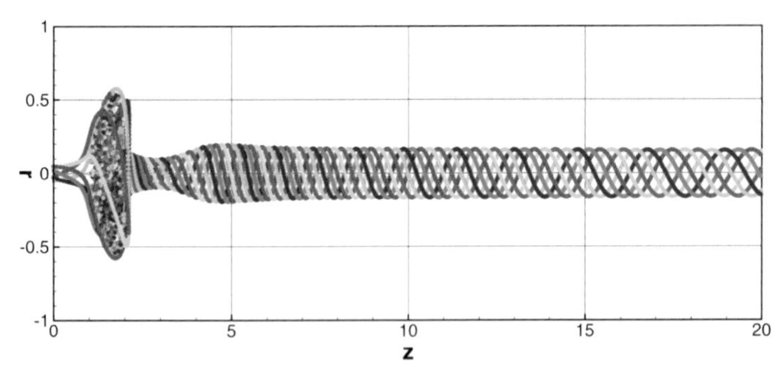

Figure 1 Swirl parameter $S = 1.1$, quasisteady state.

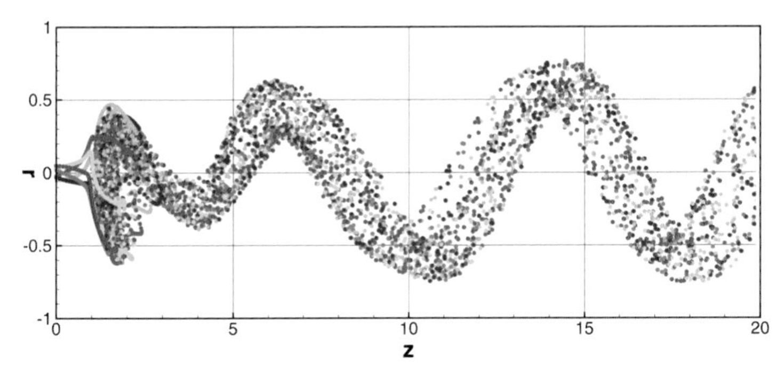

Figure 2 $S = 1.1$, single helical breakdown structure.

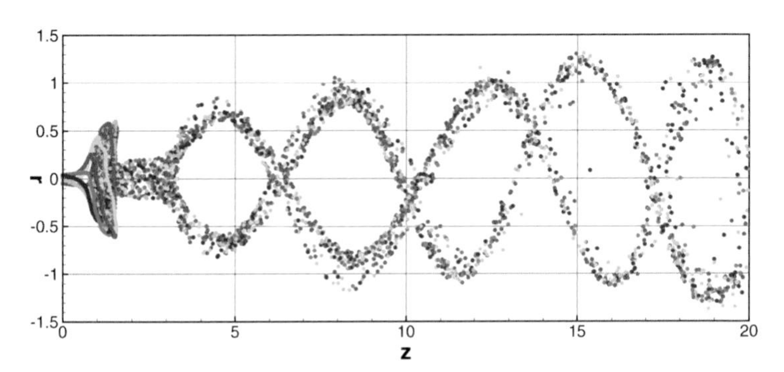

Figure 3 $S = 1.3$, double helical breakdown structure.

Breakdown modes of swirling jets with coflow

M. R. Ruith

University of Southern California

E. Meiburg

University of California, Santa Barbara

We present numerical simulations[1] of vortex breakdown modes of spatially growing, swirling jets with coflow in infinite domains at Reynolds number $\mathrm{Re} = 200$ based on the vortex core radius. A fixed inflow velocity profile[2] with uniform axial flow and an azimuthal velocity component represented by the swirl parameter S is prescribed at the left boundary. Here S represents the ratio of the azimuthal velocity at the core edge to the axial velocity at infinity. Particles are released at the inflow plane close to the axis to display streaklines of different color.

For small values of S (not shown), a stable, axisymmetric flow evolves that is characterized by a closed recirculation bubble. At larger values of S, an axisymmetric quasisteady state develops that displays a pronounced swelling in the wake of the bubble (Fig. 1). Here an azimuthal instability develops, which ultimately yields a helical breakdown (Fig. 2). Increasing the swirl number to $S = 1.3$ replaces the single helix with a double helical breakdown mode (Fig. 3).

[1] M. R. Ruith, P. Chen, E. Meiburg and T. Maxworthy, "Three-dimensional vortex breakdown in swirling jets and wakes: direct numerical simulation," *J. Fluid Mech.* **486**, 331–378 (2003).

[2] W. Grabowski and S. Berger, *J. Fluid Mech.* **75**, 525 (1976).

Keywords

simulation; streaklines.

Mushrooms and snakes: a visualization of Richtmyer–Meshkov instability

P. M. Rightley, P. Vorobieff, and R. F. Benjamin

Los Alamos National Laboratory

We visualize the Richtmyer–Meshkov instability (i.e., the impulsively driven analog of the Rayleigh–Taylor instability) of a thin layer of heavy gas (SF_6) embedded in a lighter gas (air) and accelerated by a planar shock passing through a shock tube. The heavy gas layer (or curtain) is formed by SF_6 flowing vertically downward into the test section of the shock tube through a varicose nozzle and out of the test section through a slit. Thus, there are no membranes used in producing two diffuse density interfaces on the upstream and downstream sides of the curtain. A horizontal CW laser sheet illuminates the curtain near the mid-plane of the shock tube. An intensified CCD camera records multiple exposures of the shock-accelerated curtain as the latter is swept downstream at the shock piston velocity. This mean motion in conjunction with a variable, intensifier-controlled inter-exposure time allows the acquisition of several images of the curtain on the CCD array during the course of the 1 ms event. In the images shown, the first exposure (left) depicts the pre-shock initial condition of the curtain. The timing of each subsequent exposure (after shock interaction) is shown on the images. Details of the experiment are provided elsewhere.[1]

The images show the three primary flow morphologies seen in previous studies:[2] upstream and downstream mushrooms as well as the sinuous mode. Within any given cross-stream wavelength, the morphologies correlate with the magnitude of initial perturbation of either the upstream or downstream edge of the curtain. For the first time, this study also shows coexistence of multiple morphologies during the same event. Exact reproduction of the initial conditions during the experiment is difficult due to slow variations in the shape of the curtain prior to shock impact. This study is the first to produce multiple exposures during a single event and to allow quantitative measurement of the growth rate of the instability and its eventual transition to turbulence.

We would like to thank Frank Kosel and Hadland Phototonics, Ltd. for the use of their camera. This work was performed under U.S. Department of Energy Contract No. W-7405-ENG-36.

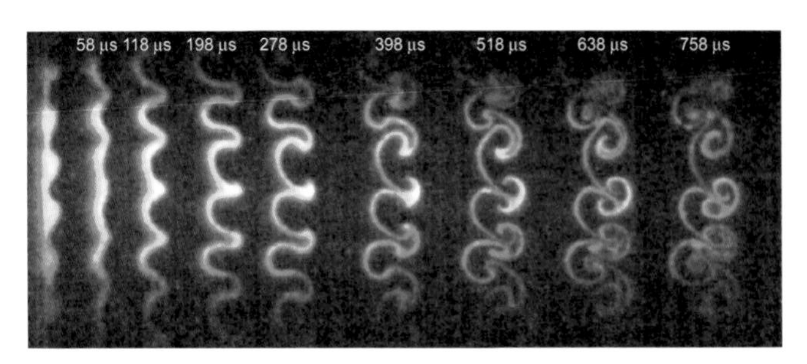

(a) Downstream mushrooms

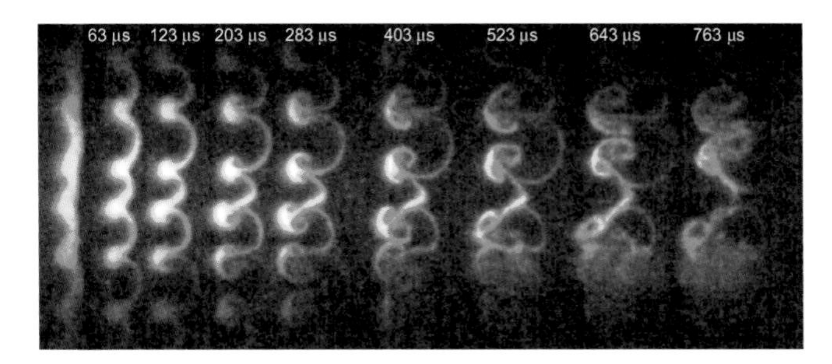

(b) Upstream mushrooms

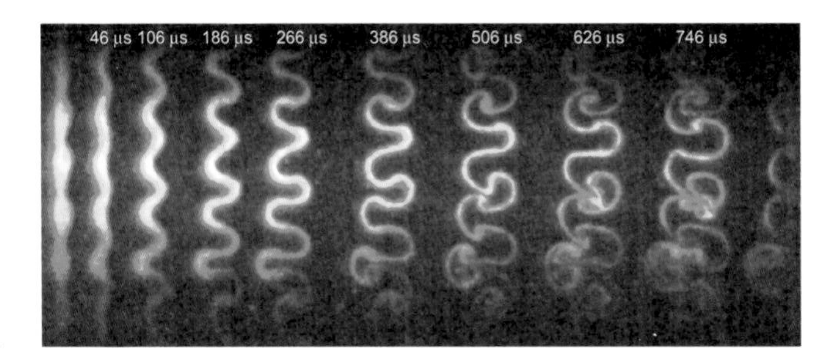

(c) Sinuous mode

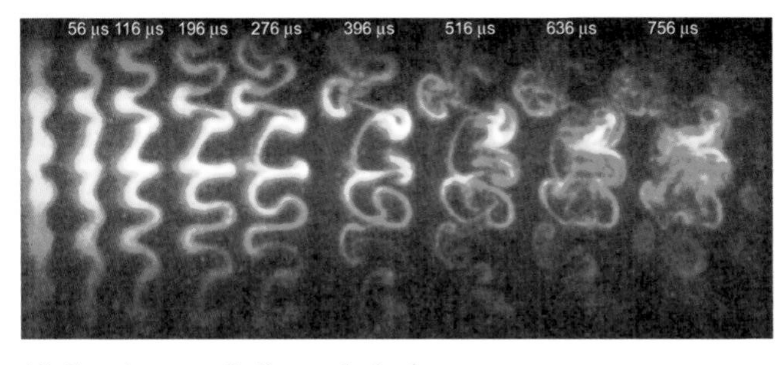

(d) Coexistence of all morphologies

Figure 1

Keywords

Rayleigh–Taylor instability; Richtmyer–Meshkov instability; shock tube; mushroom mode; sinuous mode.

[1] P. M. Rightley, P. Vorobieff, and R. F. Benjamin, "Evolution of a shock-accelerated thin fluid layer," *Phys. Fluids* **9**, 1770 (1997).

[2] J. W. Jacobs, D. G. Jenkins, D. L. Klein, and R. F. Benjamin, "Nonlinear growth of the shock-accelerated instability of a thin fluid layer," *J. Fluid Mech.* **295**, 23 (1995).

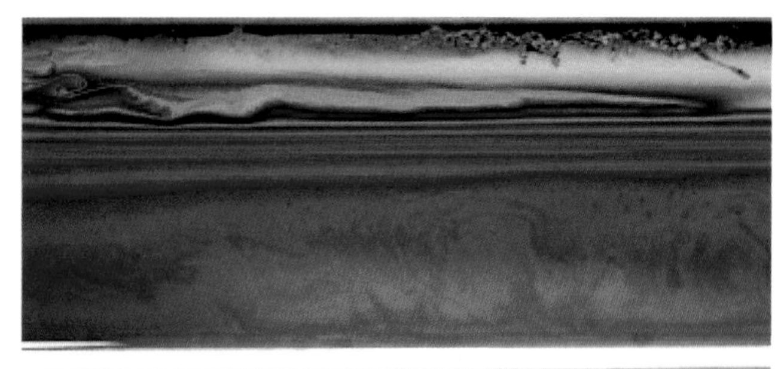

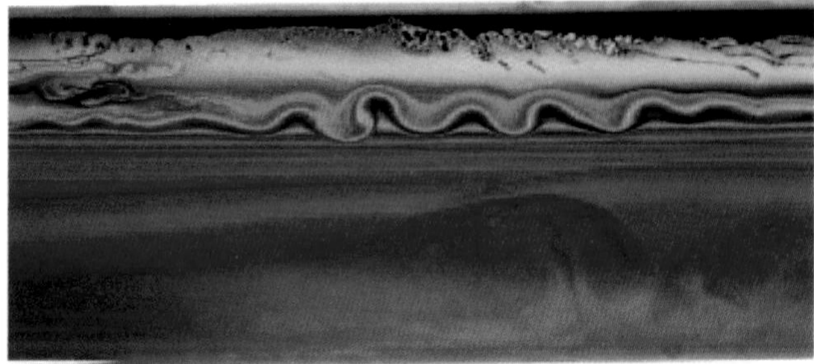

Figure 1

Evaporatively driven convection in a draining soap film

Jan M. Skotheim and John W. M. Bush
MIT

A soap film was created by dipping a rectangular wire frame of height 3.5 cm and width 15 cm into a solution of 95% by volume distilled water, 4% clear liquid Ivory soap, and 1% glycerine. The frame was mounted in a vertical position, and the film drained under the influence of gravity in an unsaturated environment. Evaporation of water increases the surface tension gradient in the thinnest (uppermost) regions of the film, thus disrupting the dynamical balance of the draining film and prompting the convective instability captured in this sequence of photographs.

A horizontal bump develops at the base of the gray film and grows in amplitude until becoming gravitationally unstable and so generating a series of sinking plumes of relatively thick film. The plumes penetrate a finite distance into the film, giving rise to a turbulent mixed layer which slowly erodes the underlying region of stably stratified film. The sequence spans approximately 10 seconds. Note the black film adjoining the wire frame at the top of the film, and the relatively weak convective motions, associated with marginal regeneration, evident near the base of the film.

Keywords
evaporation; convective instability; plumes.

Anomalous Saffman–Taylor fingering

Y. Couder

Groupe de Physique des Solides de l'Ecole Normale Supérieure, Paris, France

These three photographs show the evolution of the shape of the interface when air is injected at the center of a Hele-Shaw circular cell and forces oil to recede. The overall aspect of the pattern is similar to that first observed by Paterson. However, these photographs show that a local disturbance of the tip of one of the growing fingers can change drastically its regime of growth. The disturbance can be caused by a groove etched in the glass plates, by a thin thread stretched radially in the cell, or (as in the present photographs) by the presence of a small isolated bubble located at the tip of the finger. In all these cases the finger grows faster, its tip becomes parabolic, then it is affected by periodic side branching so that it becomes similar in shape to certain crystalline dendrites.[1]

It is a classical result that in a linear geometry the Saffman–Taylor finger tends to invade a fraction $\lambda = 0.5$ of the width of the channel. This result was interpreted recently as the selection by surface tension of a discrete set of solutions. Conversely we showed[2] that the disturbance of the tip leads in the linear geometry to a different regime of growth. The selection of the discrete set of solutions is removed and all the continuum of solutions with $\lambda < \frac{1}{2}$ can then be reached.

Keywords

Hele-Shaw cell; Saffman–Taylor instability; finger selection.

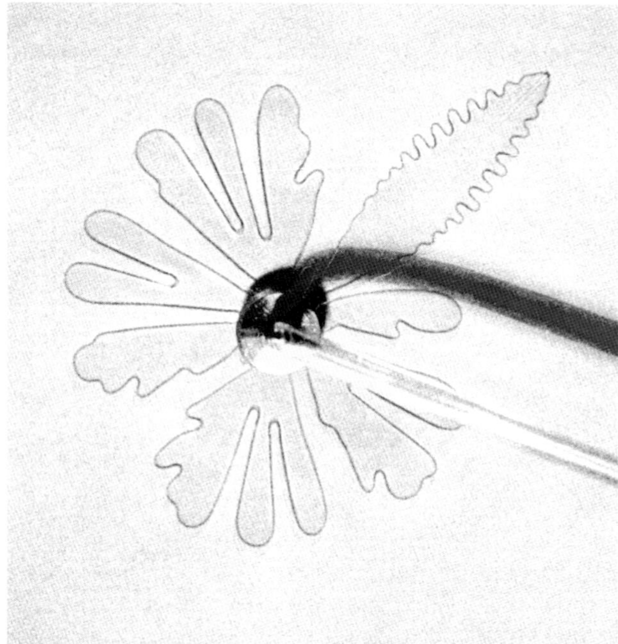

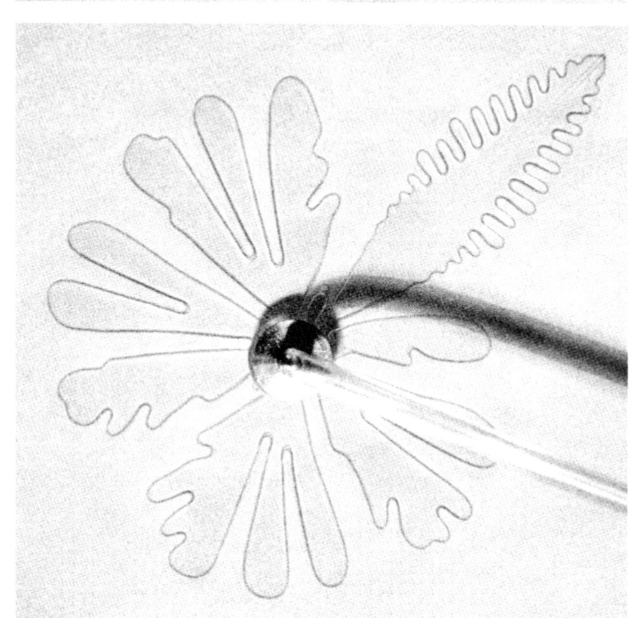

Figure 1

[1] Y. Couder, O. Cardoso, D. Dupuy, P. Tavernier, and W. Thom, *Europhys. Lett.* **2**(6), 437 (1986).

[2] Y. Couder, N. Gerard, and M. Rabaud, *Phys. Rev.* **A 34**, 5175 (1986).

Flows in the circular couette system

C. D. Andereck
Ohio State University

H. L. Swinney
The University of Texas at Austin

We have surveyed[1,2] the flow regimes for a system of independently rotating concentric cylinders, with visualization accomplished using polymer flakes (Kalliroscope AQ1000) in water.[3,4] Example flows are shown in Figs. (a) and (b). Figure (a) shows V-shaped turbulence for counter-rotating cylinders ($R_0 = -4670$, $R_i = 1070$), (b) shows a twisted vortex pattern between nearly co-rotating cylinders ($R_0 = 720$, $R_i = 1040$). The Reynolds numbers are given by $R_0 = \Omega_0 bd / \nu$ and $R_i = \Omega_i ad / \nu$, where $\Omega_i(\Omega_0)$ is the inner (outer) cylinder angular velocity, $a(b)$ is the inner (outer) cylinder radius, $d = b - a$, and ν is the kinematic viscosity. This visualization method has been used for qualitative identification of new flow regimes and quantitative measurements of wavelengths, pattern speeds, and frequencies in the flows. It has also proved invaluable as a complement to laser Doppler velocimetry.

Keywords
flake reflection; rotating cylinders; Taylor–Couette cell.

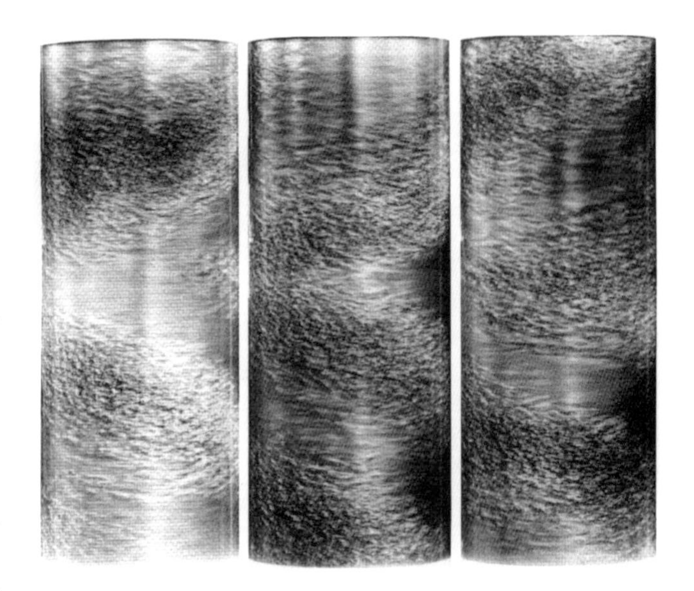

(a)

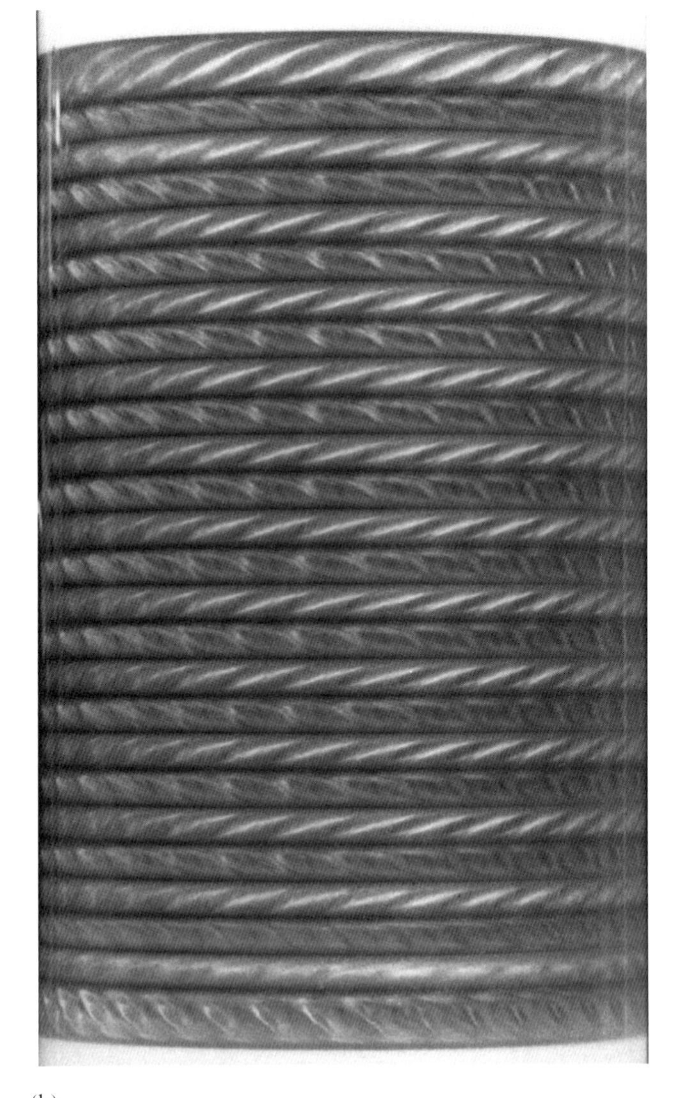

(b)

Figure 1

[1] C. D. Andereck, R. Dickman, and H. L. Swinney, *Phys. Fluids* **26**, 1395 (1983).

[2] C. D. Andereck, S. S. Liu, and H. L. Swinney, *J. Fluid Mech.* **164**, 155 (1986).

[3] P. Matisse and M. Gorman, *Phys. Fluids* **27**, 795 (1984).

[4] O. Savas, *J. Fluid Mech.* **152**, 235 (1985).

10 Transition and turbulence

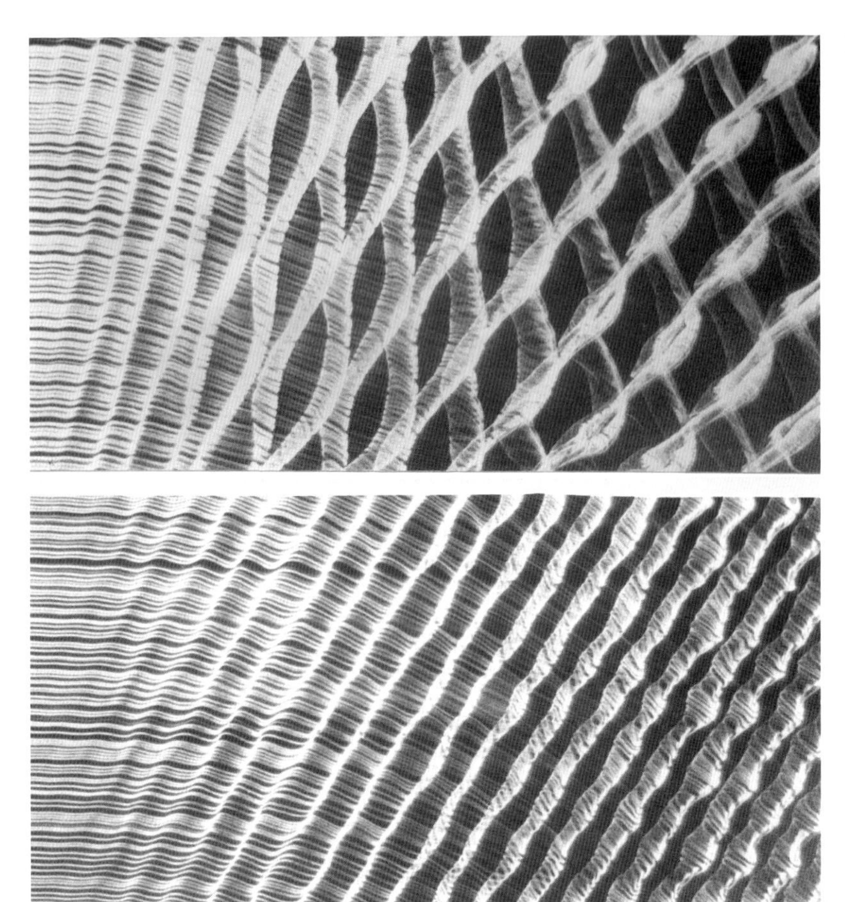

Figure 1

A new mechanism for oblique wave resonance

A. Prasad and C. H. K. Williamson

Cornell University

Despite the large body of research concerned with the near wake of a circular cylinder, the far wake, which extends beyond about 100 diameters downstream, is relatively unexplored,[1,2] especially at low Reynolds numbers. We have recently shown that the structure of the far wake is exquisitely sensitive to free-stream noise, and is precisely dependent on the frequency and scale of the near wake; indeed it is shown that the presence of extremely low-amplitude peaks in the free-stream spectrum, over a remarkably wide range of frequencies, are sufficient to trigger an "oblique wave resonance" in the far wake.[3–5]

We show, in the upper photograph of Fig. 1, a nonlinear interaction between oblique shedding waves generated from upstream (to the left) and 2-D waves amplified downstream from free-stream disturbances (in the central region). We use the "smoke-wire" technique (placed 50 diameters down-stream), and the wake is viewed in plan-view, with flow to the right. This two-wave interaction triggers a third wave, namely an "*oblique resonance wave*" at a large oblique angle, to grow through nonlinear effects (in the right half of the photograph), in *preference* to the original two waves. If smoke is introduced 100 diameters downstream, in the lower photograph (under slightly different conditions), then all that is seen is a set of such large-angle oblique resonance waves.

This work is supported by the Office of Naval Research.

Keywords

smoke visualization; far wake; shedding waves.

[1]J. M. Cimbala, H. M. Nagib, and A. Roshko, "Large structure in the far wakes of two-dimensional bluff bodies," *J. Fluid Mech.*, **190**, 265 (1988).

[2]M. Hammache and M. Gharib, "On the evolution of three-dimensionalities in laminar bluff body wakes," *Proceedings of the IUTAM Conference on Bluff Body Wake Instabilities*, ed. H. Eckelmann, J. M. R. Graham, P. Huerre, and P. A. Monkewitz (Springer-Verlag, Berlin, 1992), pp. 349–357.

[3]C. H. K. Williamson and A. Prasad, "Oblique wave interactions in the far wake," *Phys. Fluids* **A 5**, 1854 (1993).

[4]C. H. K. Williamson and A. Prasad, "A new mechanism for oblique wave resonance in the 'natural' far wake," *J. Fluid Mech.* **256**, 269 (1993).

[5]C. H. K. Williamson and A. Prasad, "Acoustic forcing of oblique wave resonance in the far wake," *J. Fluid Mech.* **256**, 315 (1993).

Visualization of different transition mechanisms

W. S. Saric
Arizona State University

The sequence of photos in Figs. 1(a)–1(d) illustrates the different types of boundary-layer transitions that occur as a function of Tollmien–Schlichting (T–S) wave amplitude and fetch.[1] The presence of these different mechanisms was first revealed through flow visualization,[2] and subsequent hot-wire measurements[1–3] confirmed the details. The distance from the leading edge is shown in cm. Tollmien–Schlichting waves are introduced by means of a vibrating ribbon located at 48 cm (near branch I of the neutral stability curve) at a frequency of 39 Hz. The flow velocity is 6.6 m/sec. An IIT-type smoke wire is placed at $x=138$ cm and $y=0.1$ cm. Branch II of the neutral stability curve for dimensionless frequency $F=83\times10^{-6}$ is $x=170$ cm. The T–S wave amplitude is referenced to the branch II point.

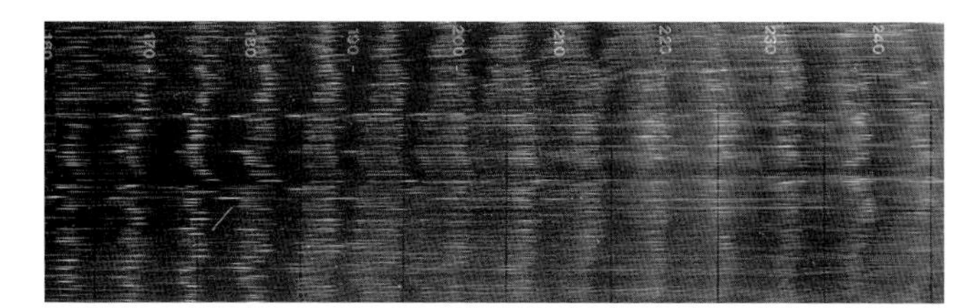

Figure 1(a)

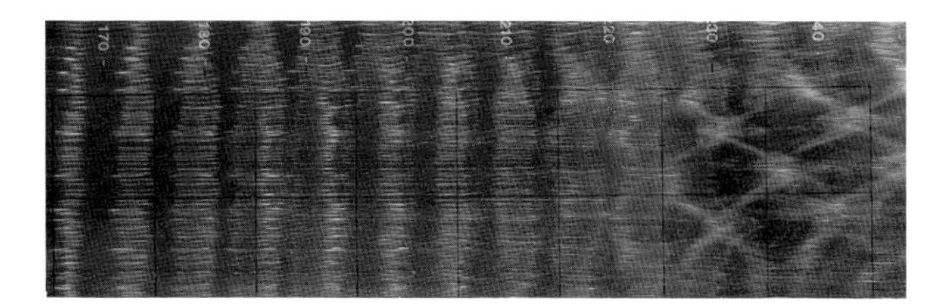

Figure 1(b)

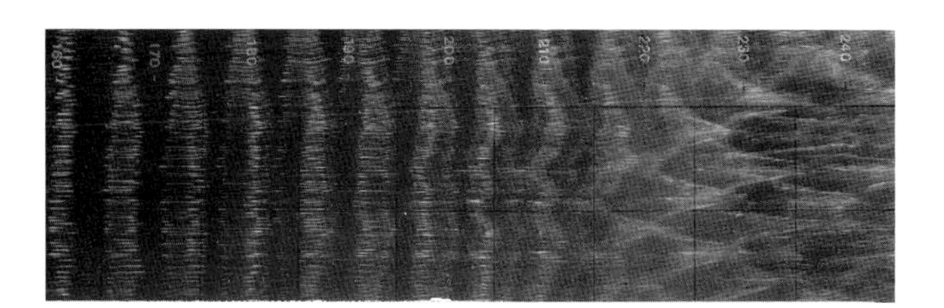

Figure 1(c)

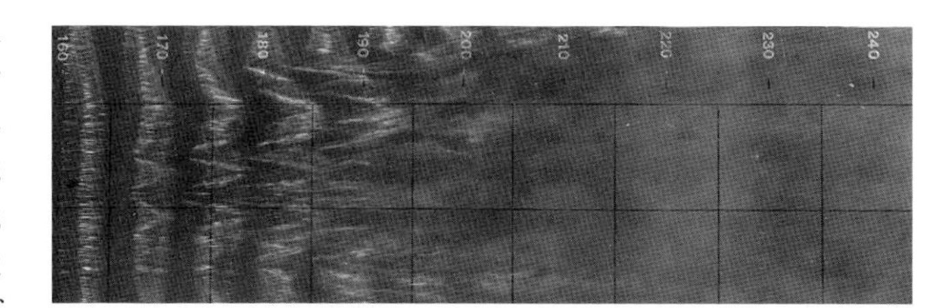

Figure 1(d)

Figure 1 shows (a) 2-D T–S waves where $u'=0.2\%$, (b) staggered structure I where $u'=0.3\%$, (c) staggered structure II where $u'=0.4\%$, (d) an ordered peak-valley structure where $u'=1\%$.

As the T–S wave amplitude is increased through Figs. 1(a)–1(d), the pattern changes from *regular* T–S waves to a *staggered* 3-D pattern with a *large* spanwise wavelength to a staggered pattern with a *small* spanwise wavelength to an *ordered* peak-valley pattern. The staggered peak-valley structure is a sufficient condition for the existence of subharmonics. Note that subharmonic breakdown occurs outside of the unstable region of the primary wave. This proved the value of the flow visualization. Reference 2 concludes that Fig. 1(b) is accounted for by the theory of Craik, while 1(c) is described by the theory of Herbert, and 1(d) is typical of the experiments of Klebanoff. Figure 2 is a closeup of the ordered peak-valley structure of the Klebanoff-type breakdown process.

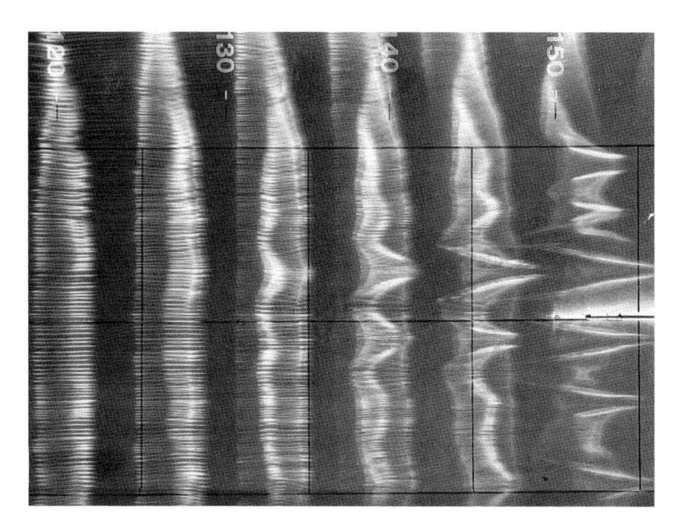

Figure 2

This work was supported by the Air Force Office of Scientific Research under Contract Nos. 82-NA-0229 and 85-NA-077.

Keywords
boundary-layer transition; Tollmien–Schlichting wave; Klebanoff breakdown; smoke visualization.

[1] W. S. Saric and A. S. W. Thomas, *Turbulence and Chaotic Phenomena in Fluids*, ed. T. Tatsumi (North-Holland, Amsterdam, 1984).
[2] A. S. W. Thomas and W. S. Saric, *Bull. Am. Phys. Soc.* **26**, 1252 (1981).
[3] W. S. Saric, V. V. Kozlov, and V. Ya. Levchenko, *AIAA* Paper No. 84-0007 (1984).

Figure 1

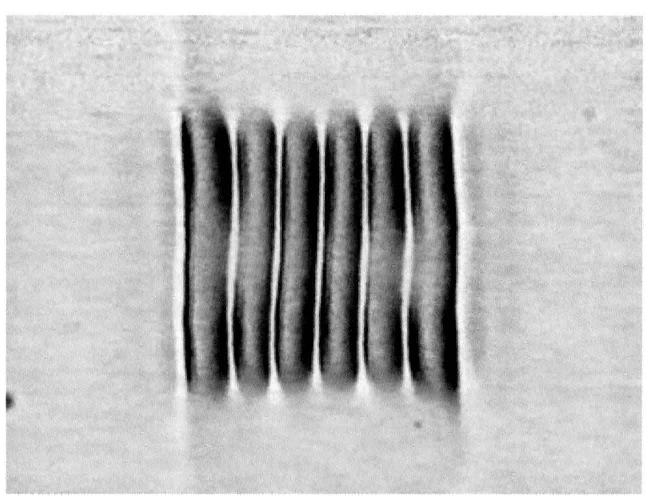

Figure 2

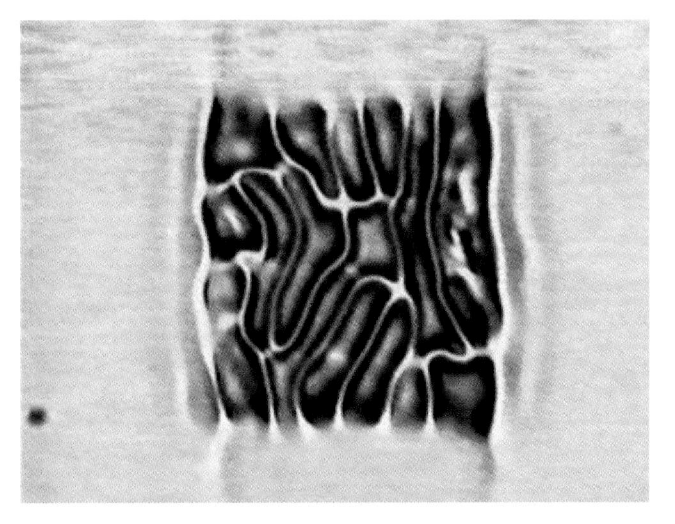

Figure 3

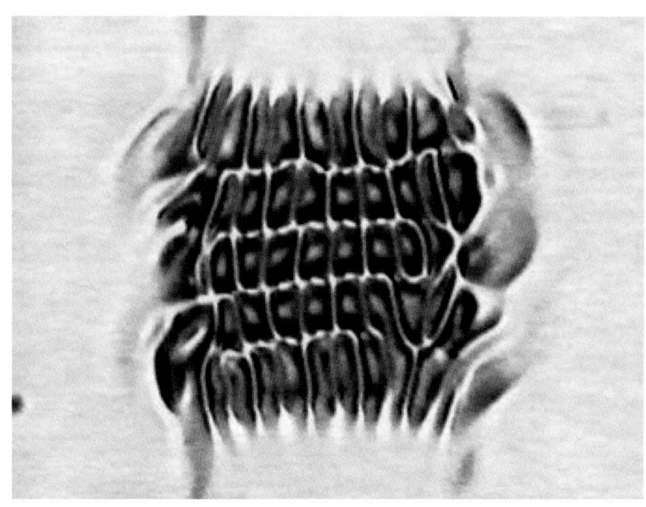

Figure 4

The transition to turbulence in a microscopic fluid flow

T. Peacock and T. Mullin

University of Manchester

The experiment comprised a 46 μm thick layer of nematic liquid crystal sandwiched between two optically flat glass plates. A line electrode of thickness 185 μm was etched onto the inner surface of each plate. The arrangement was such that, when viewed from above, the lines overlapped at right angles. This created an active region of aspect ratio 4:4:1 to which an electric field could be applied. An image of the active region is shown in Figure 1, in which a human hair has been included for scale.

As the electric field strength was smoothly increased a six roll flow developed, and this is shown in Figure 2. Associated with the convection is a spatially varying refractive index. This enables the flow to be observed directly as an intensity pattern formed by the focusing and defocusing of transmitted light. Upon further increasing the field strength, this system underwent a sequence of transitions to turbulence.[1] Two flows respectively realized in this sequence are presented in Figures 3 and 4. The former shows a weakly turbulent state and the latter shows an array of square convection cells, both of which are found in the parameter regime between "chaos" and "turbulence."

This work was supported by the EPSRC and SHARP Labs. of Europe Ltd.

[1] T. Peacock, D. J. Binks, and T. Mullin, "from low- to high-dimensional dynamics in a microscopic fluid flow," *Phys. Rev. Lett.* **82** (7), 1446 (1999).

Keywords
nematic liquid crystal; electric field; convection cells.

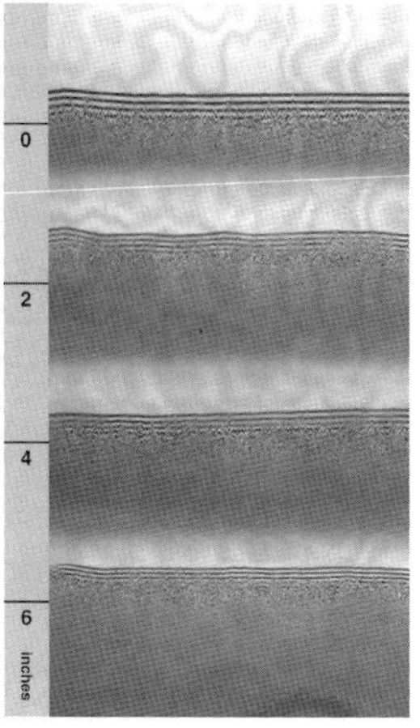

Figure 1 Laminar (no trip).

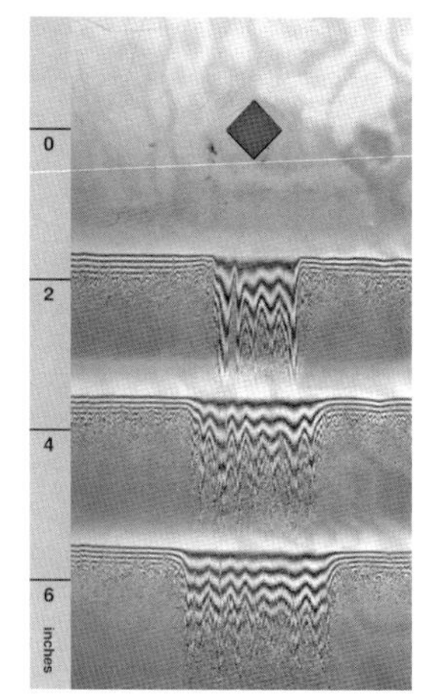

Figure 2 Diamond, $d/k=8.8$.

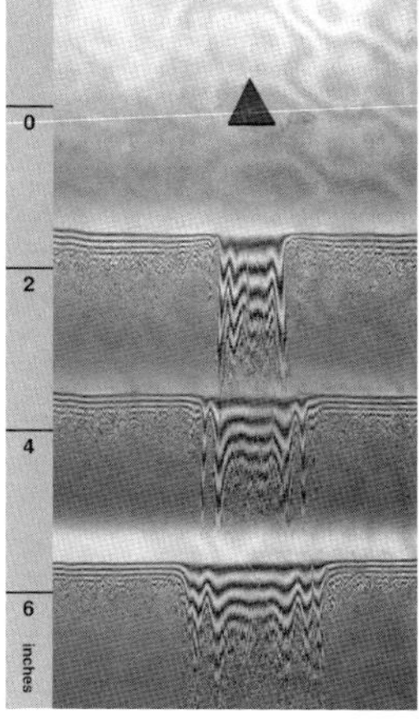

Figure 3 Triangle, $d/k=7.5$.

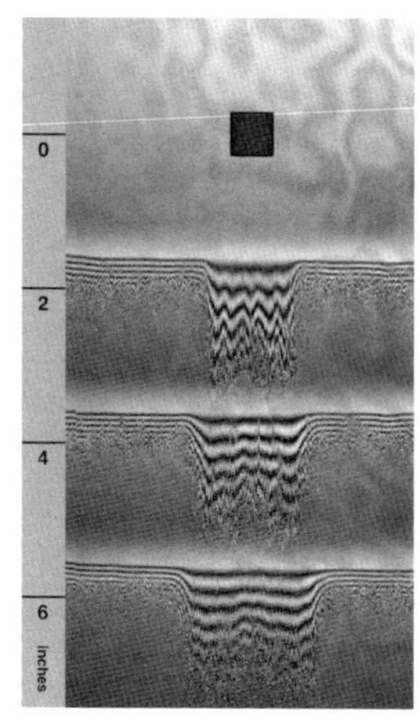

Figure 4 Square, $d/k=8.8$.

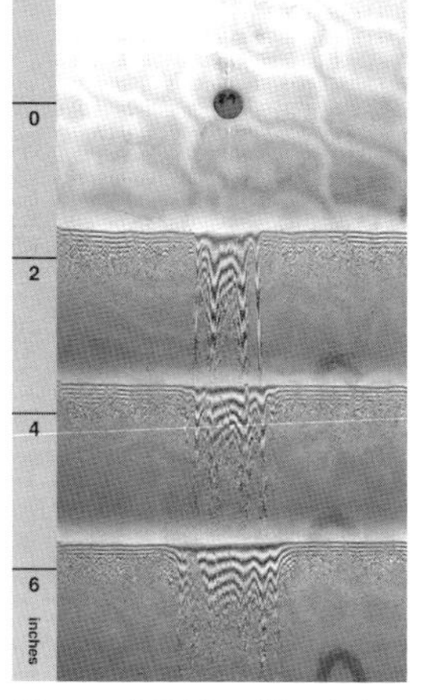

Figure 5 Disk, $d/k=4.4$.

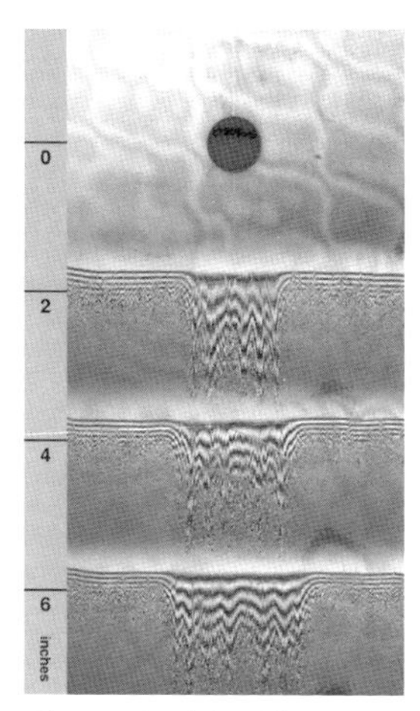

Figure 6 Disk, $d/k=8.8$.

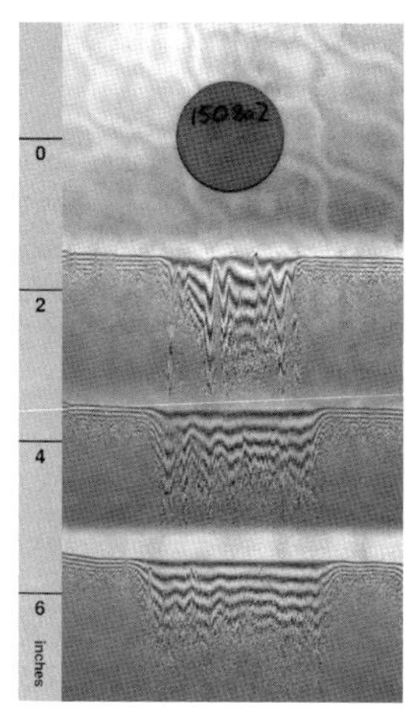

Figure 7 Disk, $d/k=17.6$.

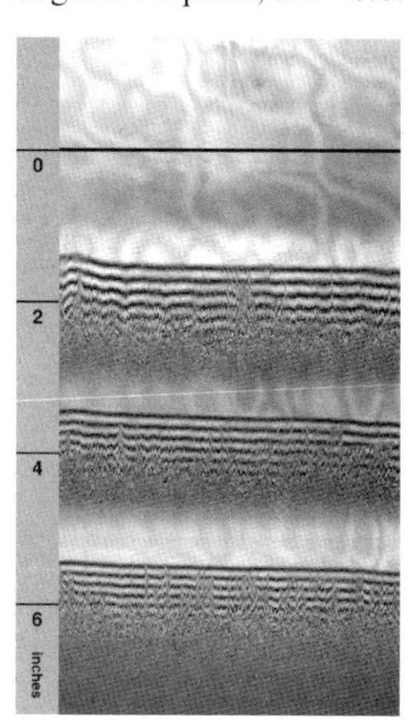

Figure 8 Wire, $Re_k=560$.

Interferometric skin friction footprint for a tripped boundary layer

Aaron Drake and Russell V. Westphal

Washington State University

Robert A. Kennelly, Jr. and David M. Driver

NASA Ames Research Center

Skin friction variations in the vicinity of isolated trip elements have been visualized with oil film interferometry. Lines of transparent silicone oil, applied perpendicular to the onset flow, were thinned to yield an oil film whose thickness distribution reflected regions of laminar and turbulent flow when visualized with interferometry. Near-normal illumination of the thinned oil film using a bank of mercury vapor lamps was used to create the interferograms, which were photographed with a conventional 35-mm camera fitted with a narrow bandpass filter. The test surface was covered with a self-adhesive film consisting of black paint-backed Mylar to yield an optically suitable surface for interferometry.

The images shown here were obtained during studies[1] of the laminar boundary layer perturbed with trip elements of different shapes. The boundary layer developed in a slight favorable pressure gradient on a 1.2 m chord plate within the 61 cm × 61 cm test section of a low-speed wind tunnel. All trip elements were located 16 cm downstream of the leading edge of the test surface, corresponding to a streamwise Reynolds number of 100,000. The height k of each of the isolated elements was about 75% of the boundary layer thickness, giving a Reynolds number based on the element height and the undisturbed velocity at the element height of $Re_k = 1240$. The laminar flow results shown in Figure 1, compared with trip elements of different shapes and widths d, all produce a downstream skin friction footprint characterized by more widely spaced interference fringes (Figures 2–7), indicative of turbulent flow. The turbulent region spreads laterally at a half-angle of 5° in each case. For comparison, results for a wire-tripped turbulent boundary layer are shown in Figure 8.

[1] A. Drake, "Effects of Cylindrical Surface Protrusions on Boundary Layer Transition," PhD thesis dissertation, Washington State University, May 1998.

Keywords

laminar-turbulent transition; boundary-layer instability.

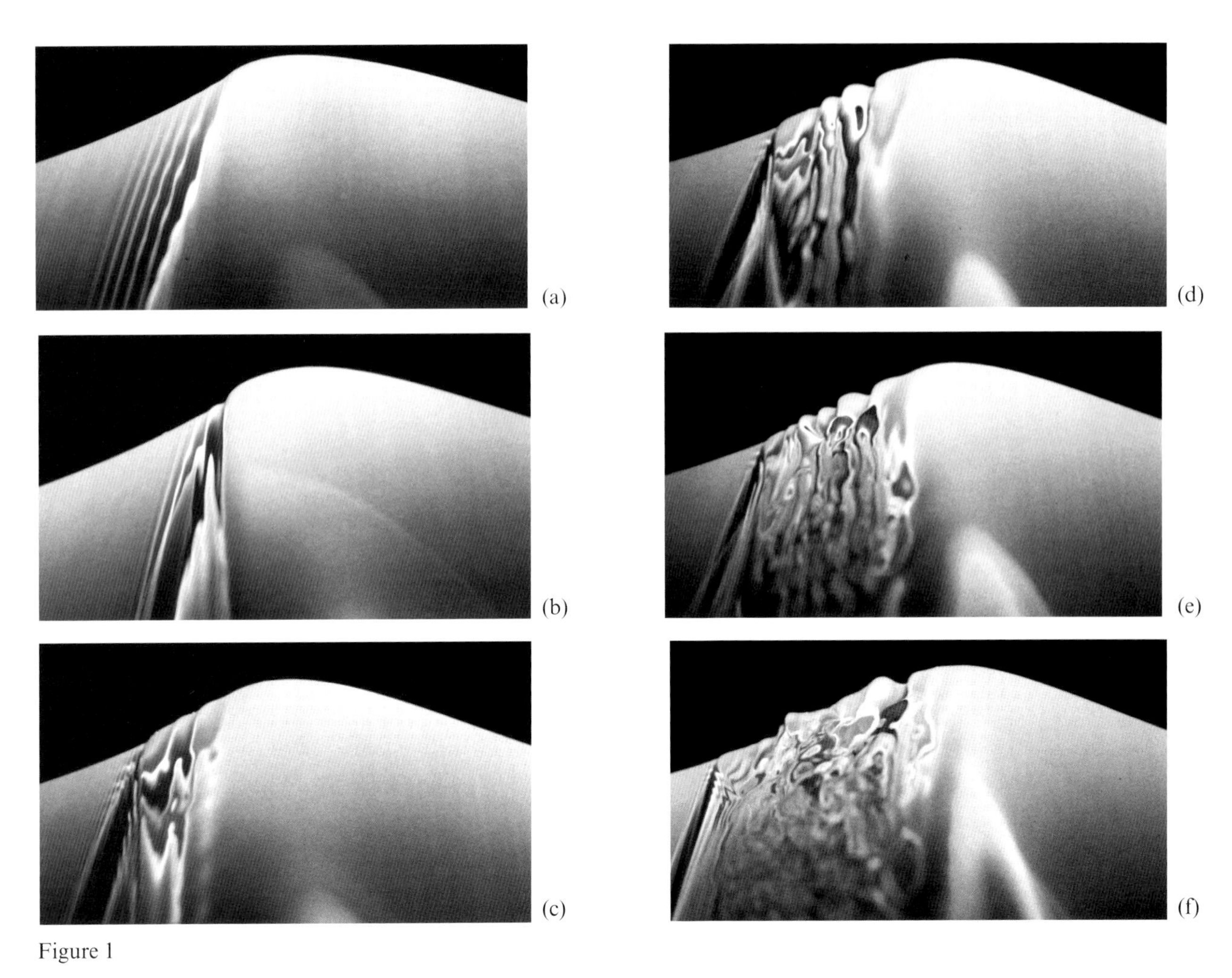

Figure 1

The formation of a spilling breaker

J. H. Duncan, V. Philomin, and H. Qiao
University of Maryland

J. Kimmel
Fachhochschule für Technik Mannheim

Photographs of an unsteady spilling breaking water wave during the transition from a laminar to a turbulent flow are presented. The camera views the wave from the side, looking down from above the water surface with an angle of about 20° from the horizontal. A laser light sheet shines vertically down and is aligned with the centerline of the tank. Fluorescent dye is mixed with the water. The dye in the light sheet glows and forms the light source for the photographs.

The breaking process starts with an asymmetric deformation of the crest of the wave such that a bulge forms on the forward face [Fig. 1 (a)]. Capillary waves are generated upstream of the toe of the bulge (see schematic). In the latter part of the breaker formation [Figs. 1 (c)–(e)], the toe of the bulge moves rapidly down the forward face of the wave. At this time a short wavelength disturbance appears in the region between the crest of the gravity wave and the toe of the bulge. This disturbance wave exhibits a very high growth rate and quickly breaks down into a random pattern indicating that the flow is turbulent.

This work was supported by the Office of Naval Research under Contract No. N00014-90-J-1977.

Keywords
water wave; wave breaking; capillary wave; gravity wave; fluorescent dye.

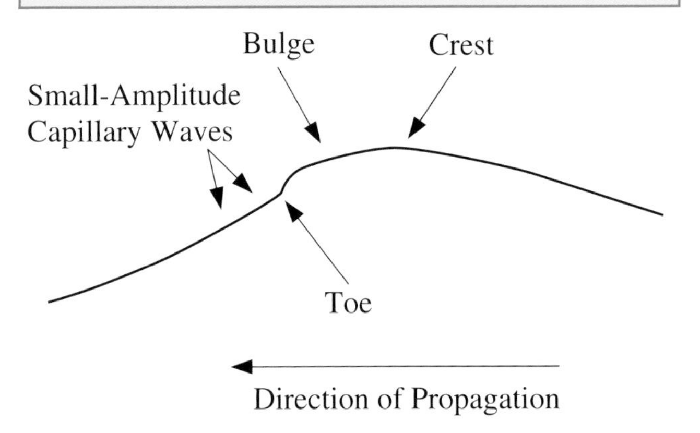

Figure 2

Figure 1

Figure 2

Figure 3

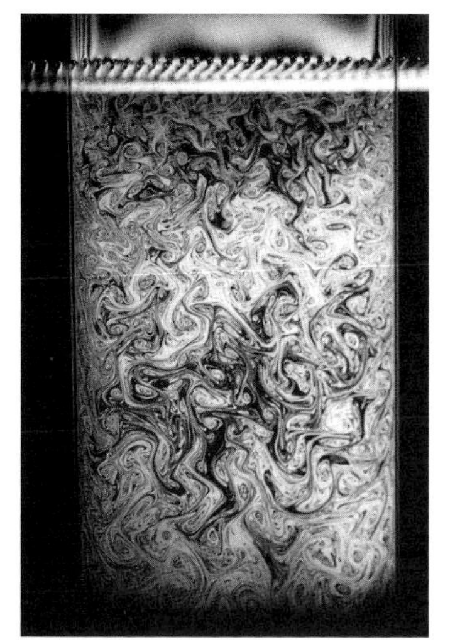

Figure 4

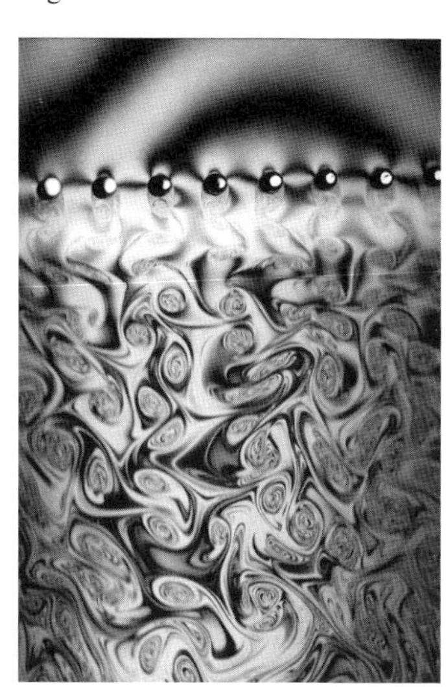

Figure 5

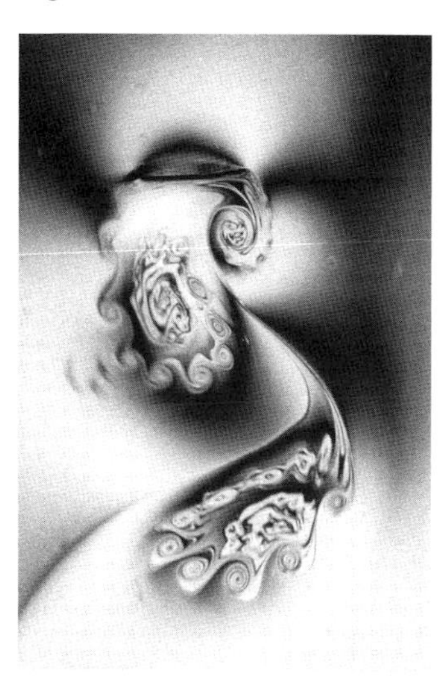

Figure 6

The onset of two-dimensional grid generated turbulence in flowing soap films

Maarten A. Rutgers, Xiao-lun Wu, and Walter I. Goldburg

Department of Physics and Astronomy, University of Pittsburgh

A free standing soap film, bounded by two vertical nylon wires, originates about 1 m above the photographed area from a bottle of soap solution. In the photographs the film has reached a terminal velocity between 2 and 3 meters per second and has a nearly uniform thickness. At the top of the photographs a comb punctures the film. The vortices generated in the wake of the comb's teeth mingle to varying degrees depending on the Reynolds number of the flow. The comb tooth diameter is 0.12 cm. The center-to-center tooth spacing is 0.3 cm. The channel is about 4 cm wide. A more detailed description of the apparatus and its possibilities is given by Kellay *et al.*[1]

Visualization of the flow patterns is spontaneous. The film is slightly compressed as it hits the comb teeth and perhaps slightly expanded behind the teeth. The resulting thickness variations in the film are visible due to constructive and destructive interference of light. Once created, thickness variations are mostly convected by the flow as a passive scalar. Figures 1–4 show the flow patterns for increasing Reynolds number. Figure 5 is a closeup near the comb teeth with a Reynolds number exceeding that of all other figures. Figure 6 shows the wake of a knife edge puncturing the film. The sharp edges produce additional fine structure that may be desirable in generating turbulence.

[1] H. Kellay, X.-l. Wu, and W. I. Goldburg, "Experiments with turbulent soap films," *Phys. Rev. Lett.* **74**, 3975 (1995).

Keywords

soap film; vortex generation; comb teeth.

Figure 1(a)

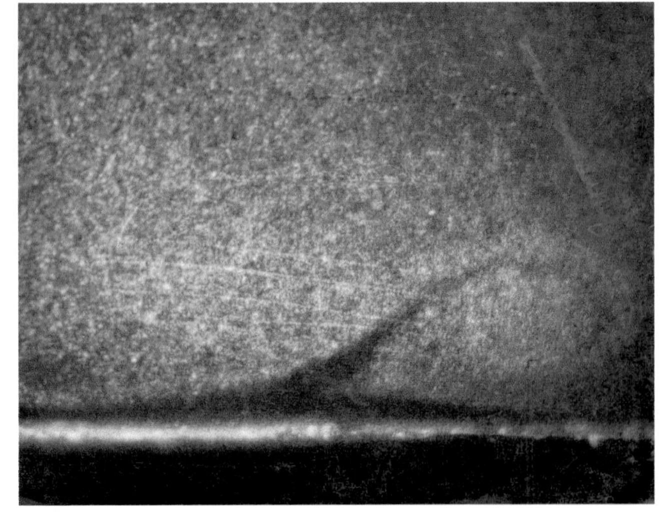

Figure 1(c)

Figure 1(b)

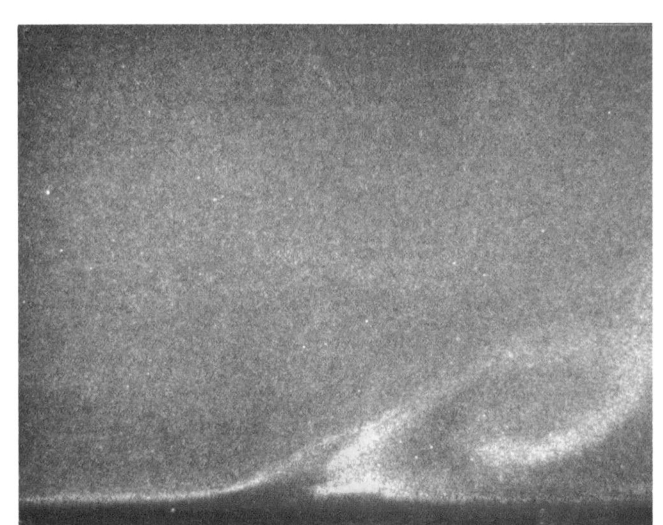

Figure 1(d)

Coherent structures in thermal turbulence

Giovanni Zocchi, Elisha Moses, and Albert Libchaber

The University of Chicago

Swirls and plumes are injected from the top thermal boundary layer in turbulent Rayleigh–Bénard convection.[1] Visualizations are made using thermochromic liquid crystals in water (Prandtl number 6) at a Rayleigh number of 10^9. The cell is a 1:1:1 box with sides of length 18.5 cm, the field of view is about 2×2 cm, and the images are rotated by 180° for clarity.

Keywords

swirl; plume; Rayleigh–Bénard convection; thermochromic liquid crystals.

[1] G. Zocchi, E. Moses, and A. Libchaber, *Physica* **A 166**, 387 (1990).

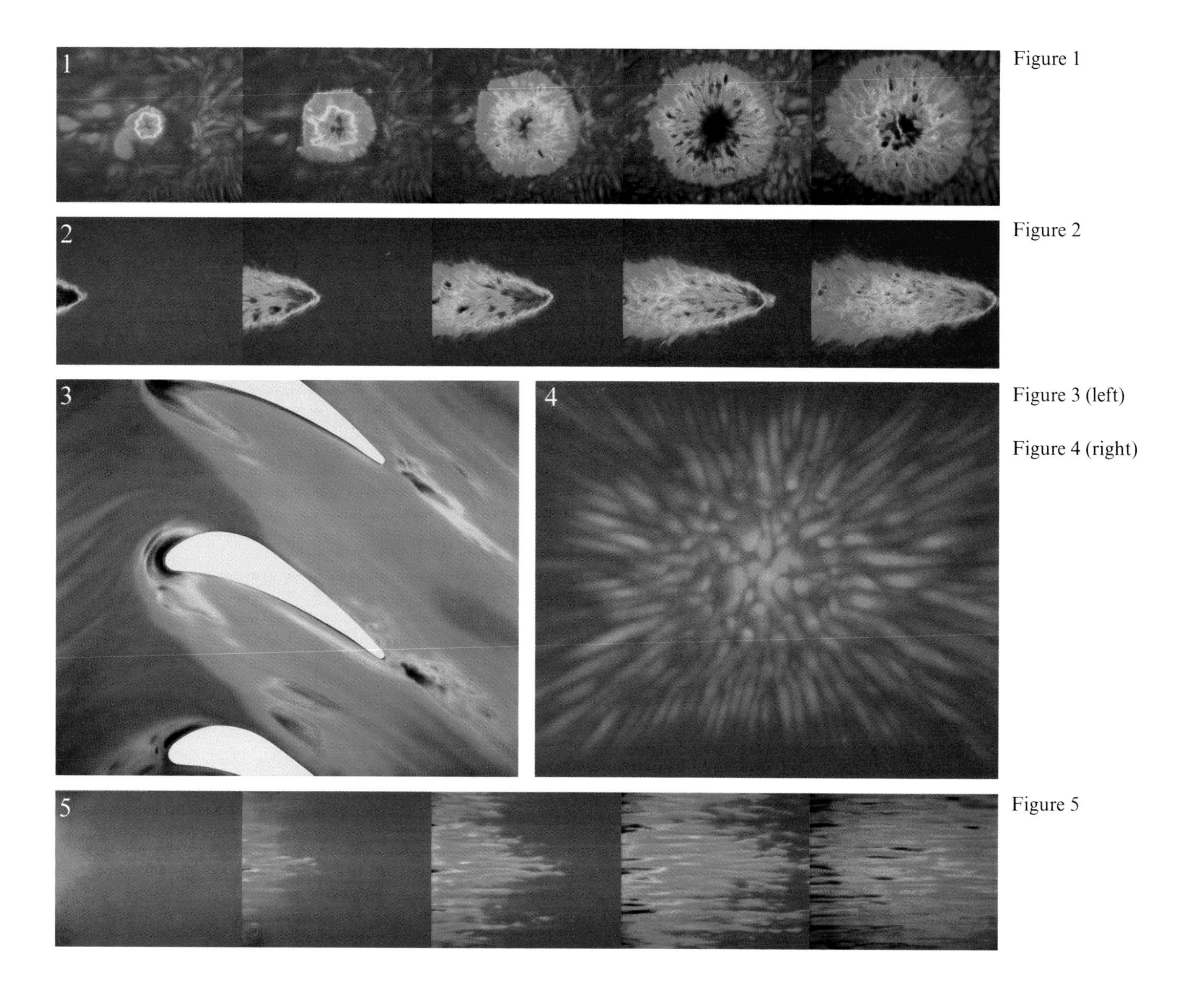

Figure 1

Figure 2

Figure 3 (left)

Figure 4 (right)

Figure 5

The colors of turbulence

D. R. Sabatino and T. J. Praisner
Lehigh University

Thermochromic Liquid Crystals (TLCs) possess unique physical properties which make them a powerful tool for temperature visualizations/measurements. The images above illustrate a variety of turbulent flows for which surface heat transfer is reflected by the color change of TLCs applied to a unique thin-film constant heat flux surface. All of the images display color patterns which are proportional to the instantaneous convective coefficients.

Figure 1 illustrates the surface heat transfer patterns for a jet of cool fluid impinging onto a warm surface. Figure 2 shows a temporal sequence of patterns generated by a passing turbulent jet of cool fluid. Instantaneous endwall temperature distributions at the base of a turbine cascade are illustrated in Figure 3. Figure 4 illustrates patterns created by free-convection cells distorted by a sink above the center of the heated plate. Finally, the sequence shown in Figure 5 shows the transition of a laminar to a fully turbulent boundary layer, as illustrated by the development of the classic low-speed streak patterns.

While TLC thermography is a useful visualization technique, it is even more valuable as a quantitative sensor of surface heat transfer. Employing a unique experimental apparatus, TLC measurements have been simultaneously combined with high resolution PIV to yield quasi-three-dimensional results. To see details of this technique and selected results, visit www.lehigh.edu/fluid.

Keywords

thermochromic liquid crystals; turbine cascade; heat-transfer patterns.

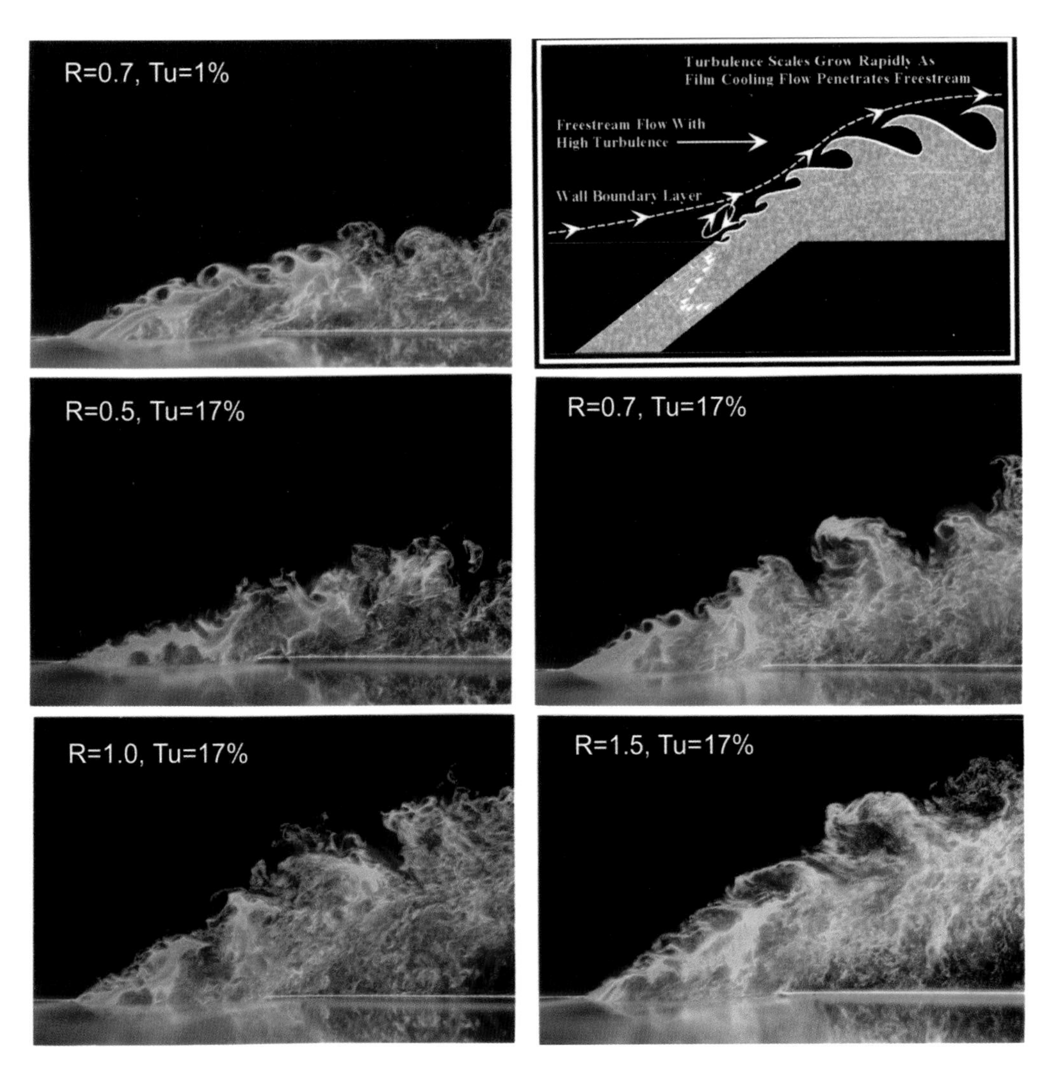

Figure 1

High free-stream turbulence influence on turbine film cooling flows

S. Gogineni
Innovative Scientific Solutions, Inc.

R. Rivir
Air Force Research Laboratory

D. Pestian
University of Dayton Research Institute

and L. Goss
Innovative Scientific Solutions, Inc.

Double pulsed two-color particle image velocimetry (PIV) images of simulated turbine film cooling flows are shown for a range of film cooling blowing ratios ($R=\rho_c U_c / \rho_\infty U_\infty$) of 0.5, 0.7, 1.0 and 1.5. The simulated turbine conditions include the film cooling jet $l/d=3$, film jet Reynolds number of 20,000, and free-stream turbulence level of up to 17%. The images are obtained by seeding only the jet flow with submicron-size smoke particles, illuminating them with a laser light sheet, and recording them on a two-color CCD sensor. These images illustrate how the jet spreads and shear layer grows with two of the problem's parameters, the blowing ratio and the free-stream turbulence level. There is a decrease in film cooling effectiveness and increased heat transfer associated with the increase in turbulence intensity which is currently difficult to predict. The PIV images and the reduced PIV data are useful in providing additional physics on mixing and dissipation for improved modeling of these flows.

Keywords
two-color PIV; film jet; shear layer.

Turbulent thermal convection over a rough surface

Yi-bing Du and Penger Tong
Oklahoma State University

In these streak pictures of small seed thermochromic liquid crystal spheres, we show visualization of the temperature and velocity fields near the upper (cold) surface of an aspect-ratio-one Rayleigh–Bénard convection cell filled with water. Cooler regions of the fluid appear brown and warmer regions appear green and blue. Two cylindrical cells are used in the experiment; one has smooth upper and lower surfaces and the other has rough upper and lower surfaces. The rough surfaces are made from identical brass plates but have woven V-shaped grooves on them. The spacing between the grooves is such that a square lattice of pyramids of 9 mm in height is formed on the surface. Figures 1 and 2 show how a cold plume erupts from the upper smooth and rough surfaces, respectively, at the Rayleigh number 2.6×10^9. In the smooth cell, the plume erupts by its own buoyancy, which accelerates the fluid in the central stem of the plume and produces a vortex ring around the stem. A two-dimensional projection of the vortex ring shows a pair of vortices with opposite signs, which gives the characteristic mushroom shape of the thermal plume.

In the rough cell, however, we find that the large-scale motion (from right to left) is modulated by the rough surface and produces an adverse pressure gradient in the groove region. This pressure gradient creates eddies whose vorticity is opposite to that of the large-scale circulation. As shown in Fig. 4, the interaction between the large-scale flow and the eddy causes the thermal boundary layer to detach near the top of the pyramid. Because the detachment of the boundary layer is driven by the large-scale flow, instead of the buoyancy force, the vortex ring generated by the buoyancy acceleration disappears and thus the thermal plumes in the rough cell lose their mushroom cap. Occasionally, we do see a few mushroom-shaped plumes in the rough cell, such as that shown in Fig. 3, but the majority of the thermal plumes look like that shown in Fig. 2. The experiment reveals that the interaction between the large-scale flow and the small eddies trapped inside the groove enhances the detachment of the thermal plumes. These extra plumes are responsible for the enhanced heat transport observed in the rough cell.[1,2]

Keywords

Rayleigh–Bénard convection; thermochromic liquid crystals; thermal plumes; boundary-layer detachment.

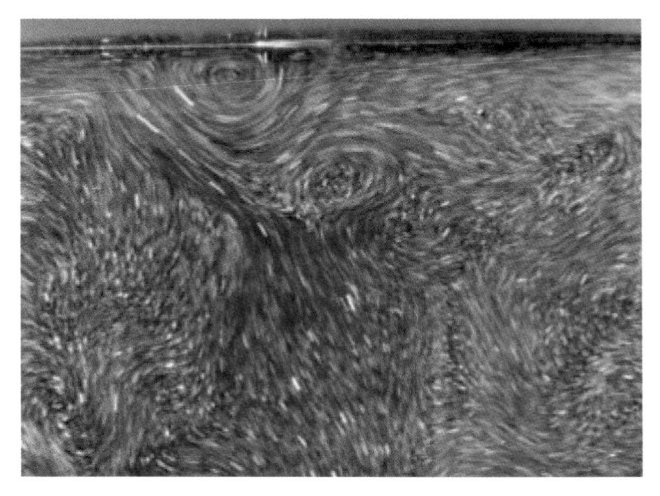

Figure 1

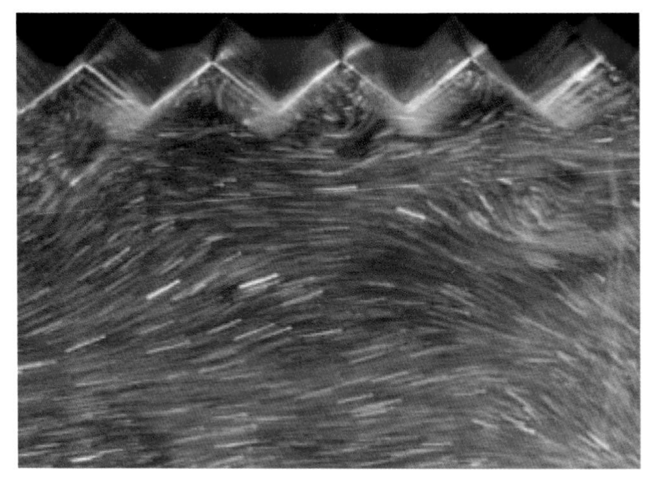

Figure 2

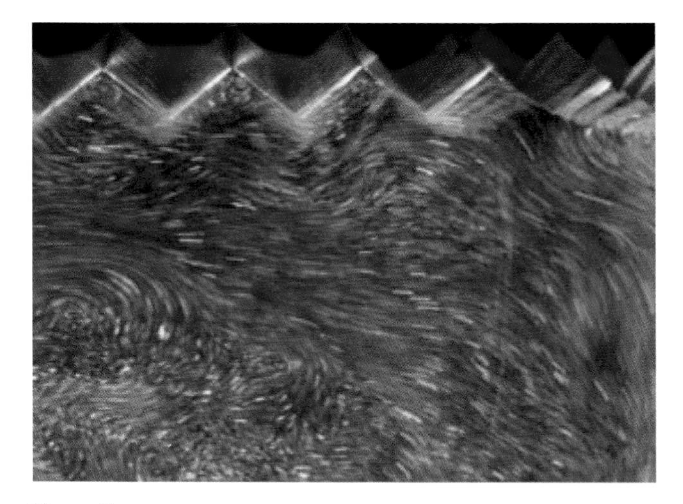

Figure 3

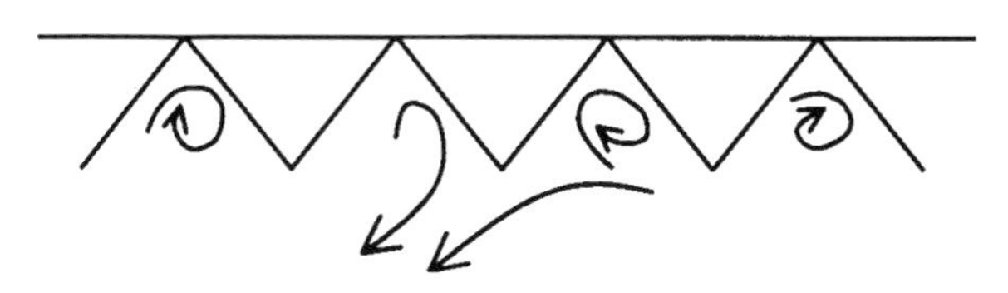

Figure 4

[1] Y.-B. Du and P. Tong, "Enhanced heat transport in turbulent thermal convection over a rough surface," *Phys. Rev. Lett.* **81**, 987 (1998).

[2] Y.-B. Du and P. Tong, "Turbulent thermal convection in a cell with ordered rough boundaries," *J. Fluid Mech.* **407**, 57 (2000).

11 Compressible flows

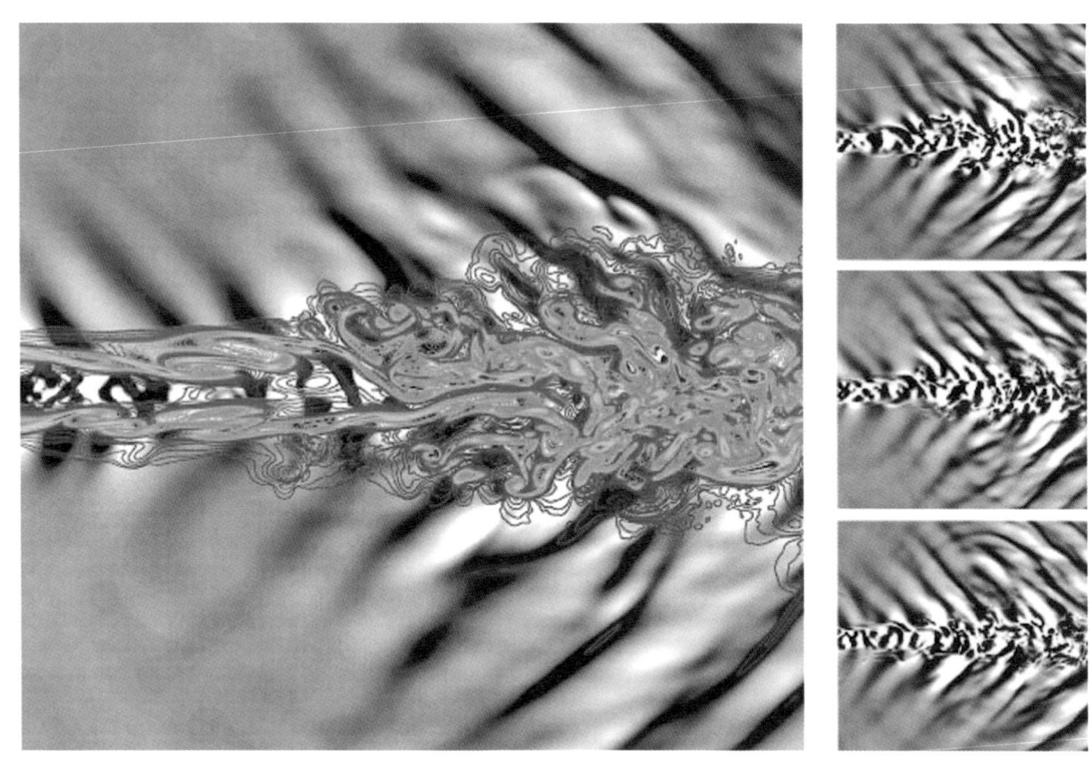

Figure 1

Mach wave radiation from a jet at mach 1.92

R. Darke and J. B. Freund
University of California, Los Angeles

The visualization shows a two-dimensional slice through the center of a Mach 1.92, Reynolds number 2000 round jet computed by direct numerical simulation. The simulation is discussed in detail by Freund, Lele, and Moin.[1] The jet turbulence is visualized with color contours of vorticity between $0U_c/r_o$ and $4.5U_c/r_o$, where U_c is the local centerline velocity and r_o is the nozzle radius. (The peak vorticity in this plane was $7.2U_c/r_o$ at the instant shown.) Despite the low Reynolds number, we see a range of turbulence scales in the flow. The grayscale levels shown are directly proportional to divergence of velocity between solid white which indicates $\nabla\cdot\mathbf{u}>0.017U_j/r_o$ and solid black which indicates $\nabla\cdot\mathbf{u}<0.017U_j/r_o$ where U_j is the jet nozzle velocity. In the sound field, narrow regions of compressions (dark) are typically separated by broader expansions (light) which suggests a nonlinear steepening process. Steepening in the far-field was investigated for this flow using weak shock theory.[1] The small visualizations, however, reveal that the dark regions appear to originate as already thin areas within the jet, as speculated by Ffowcs Williams *et al.*[2] who did not observe a significant increase in wave steepness at greater distances from a jet. This suggests a nonlinearity in the noise generation mechanism, though a quantitative investigation of this is necessary before any conclusions can be made. The role of nonlinearity in generating noise is being investigated.[3] The Mach waves all propagate at approximately 50° from the jet axis. Simultaneous visualization of the pressure and divergence of velocity (not shown) suggest that there are usually regions of high pressure at the origin of each Mach wave convecting at an appropriate supersonic velocity for the Mach angle, but as yet we have not identified consistent concrete events that spawn Mach waves.

The original simulation was conducted for the most part at Stanford University, where J. B. F. was supported by the Franklin P. and Caroline M. Johnson graduate fellowship.

[1] J. B. Freund, S. K. Lele and P. Moin, "Direct numerical simulation of a Mach 1.92 turbulent jet and its sound field," *AIAA J.* **38**, 2023 (2000).

[2] J. E. Ffowcs Williams, J. Simson, and V. J. Virchis, "Crackle: An annoying component of jet noise," *J. Fluid Mech.* **71**, 251 (1975).

[3] K. Mohseni, T. Colonius, J. B. Freund, "An evaluation of linear instability waves as sources of sound in a supersonic turbulent jet," *Phys. Fluids* **14** (10), 3593–3560 (2002).

Keywords
simulation; round jet; noise generation.

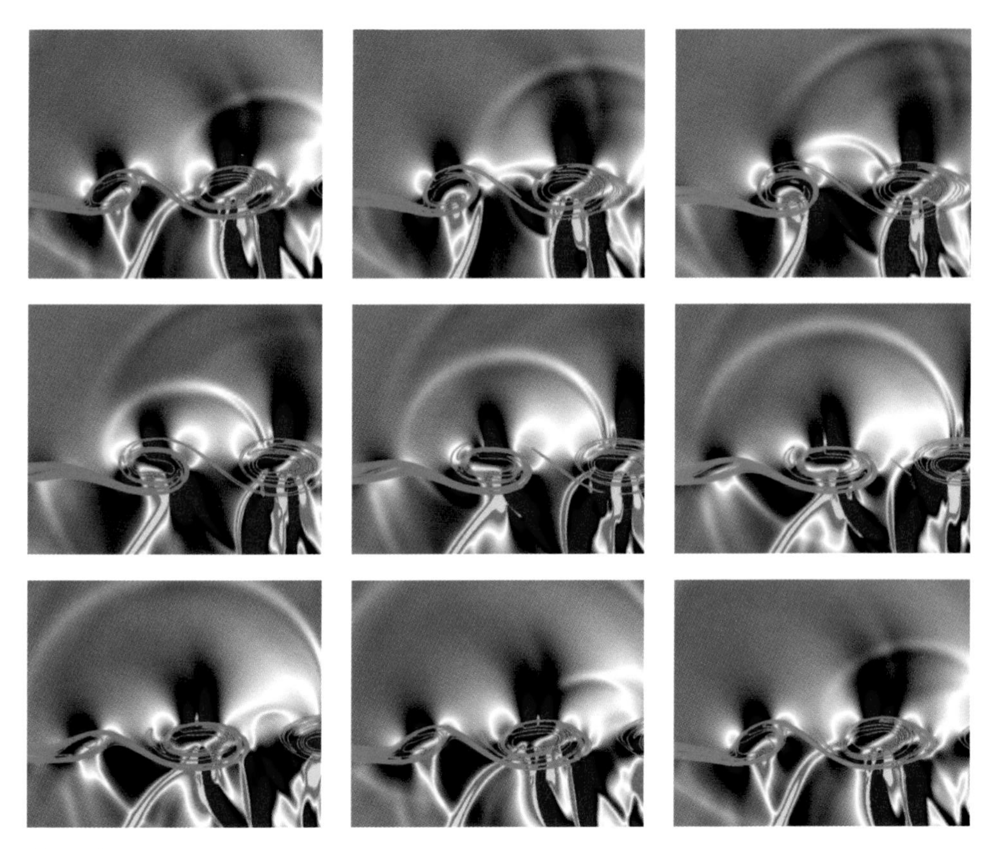

Figure 1

Sound due to the interaction of shear-layer instability-waves and an oblique shock

Ted A. Manning and Sanjiva K. Lele
Stanford University

A large component of noise from supersonic jet flows is due to the interaction of the jet's shear layer instability waves and the shock-cell structure within the jet. Supersonic jet screech is a special case of this "shock noise," whereby the radiated sound reinvigorates the instability waves near the jet nozzle and creates a positive feedback loop. Under these conditions, the amplitude of the instability waves grows rapidly to form large vortices, and the radiated sound is narrow-banded and very loud.

In an effort to understand this sound generation process, we have constructed a model problem consisting of an oblique shock impinging upon supersonic shear layer. Instability waves are imparted to the shear layer using the unstable eigenmodes at the inflow. The visualization above is the result of a two-dimensional viscous numerical simulation of such a flow. The sequence of images (starting in the upper left, moving to the right) contains the dilatation field (yellow: strong compression, red: compression, gray scale: acoustic levels, blue: expansion) overlaid by contours of vorticity (in green) for one oscillation cycle. The flow below the shear layer is supersonic ($M = 1.2$), whereas the upper stream is quiescent. The shear layer Reynolds number based on initial vorticity thickness is 1000. The shock is approximated with a low amplitude compression wave, rising 5% above the mean pressure.

The visualization indicates that the sound generation process at these large instability amplitudes can be characterized as "leakage" of the shock through the shear layer between the vortices. The thin compression front of the acoustic wave is released as the shock travels upstream; at certain points the two are continuously connected. A system of traveling waves is present in the supersonic flow downstream of the shock; these are generated during the formation of the vortex and its interaction with the shock. These also interact with the shear layer vortices to produce sound waves.

The authors would like to acknowledge the Air Force Office of Scientific Research for their support of this work under Grant No. F49620-98-10355.

Keywords
simulation; shock noise; acoustic wave

Figure 1

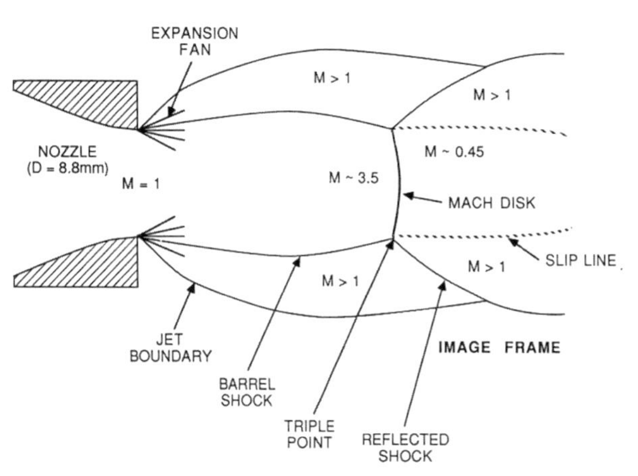

Figure 2

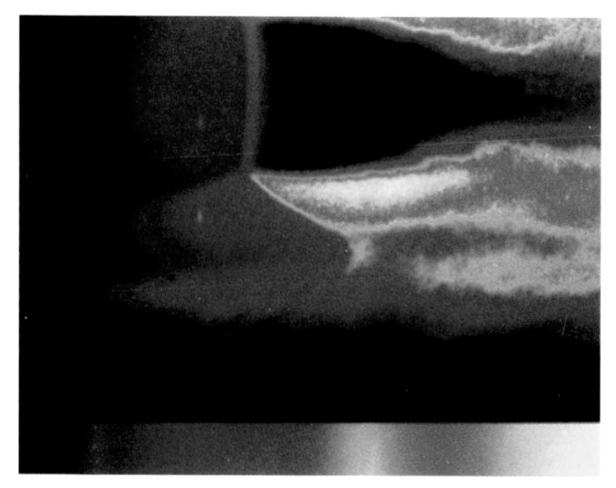

Figure 3

Visualization of a supersonic underexpanded jet by planar Rayleigh scattering

B. Yip, K. Lyons, and M. Long
Yale University

M. G. Mungal
Stanford University

R. Barlow and R. Dibble
Sandia National Laboratories

Figure 1 shows the instantaneous flow field of an underexpanded air jet (pressure ratio 7) visualized using two-dimensional Rayleigh-scattering imaging. Notable features, including Mach disk, barrel shock, reflected shock, and triple point, are labeled in Fig. 2. These features are clearly visible in the instantaneous image and in the 30-frame average shown in Fig. 3. The instantaneous image resolves the large-scale structure of the turbulent shear layer at the jet boundary and of the shear layer along the slip line downstream of the triple point. The initial growth of the shear layer along the slip line is slow because the convective Mach number is high, and compressibility effects dampen amplification of the Kelvin–Helmholtz instability. Growth rate increases with streamwise distance, as convective Mach number decreases. The size of the well-defined "large-scale" vortical structure identified in the instantaneous image is 1 mm.

Images were obtained using a pulsed, doubled Nd: YAG laser (10 nsec pulse duration) and an unintensified CCD camera with 384×576 pixels. Each pixel corresponds to 40 μm in physical space.

Keywords

Rayleigh scattering; Mach disk; barrel shock; reflected shock; shock triple point.

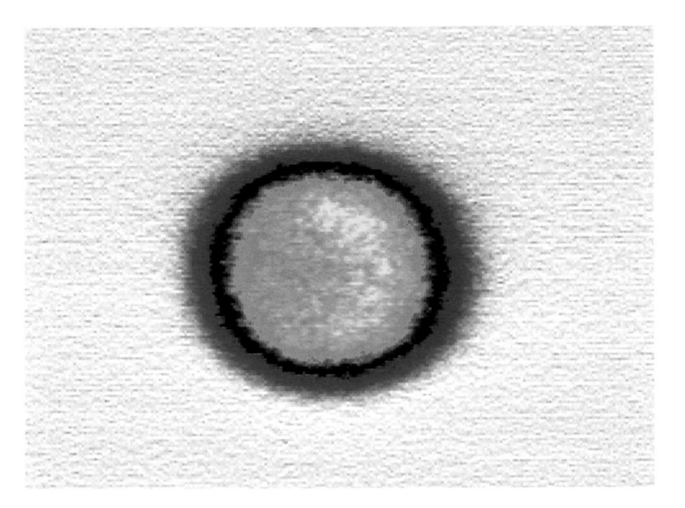

Figure 1(a)

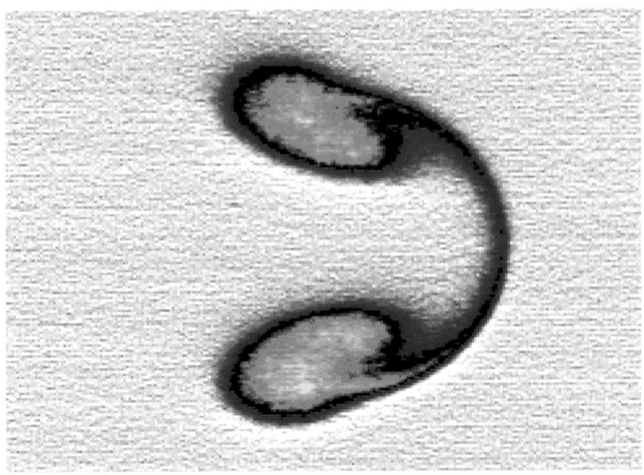

Figure 1(b)

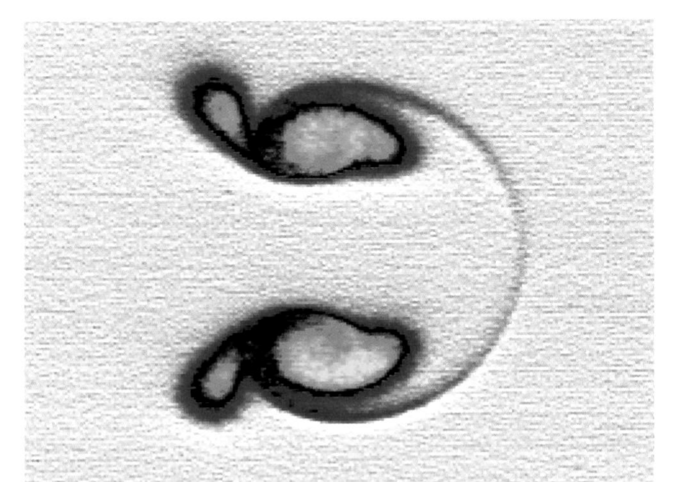

Figure 1(c)

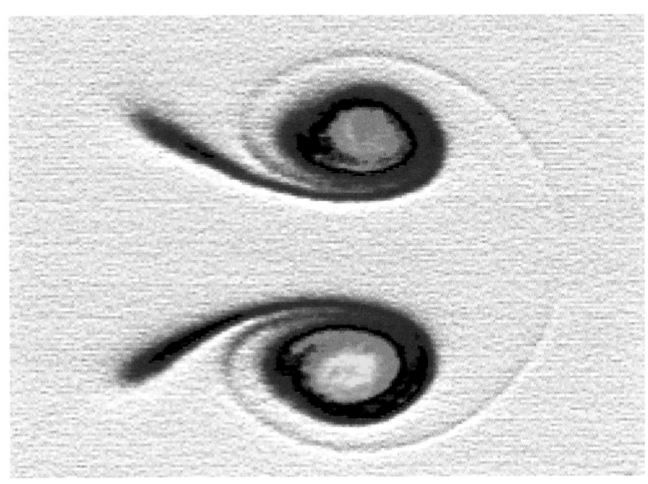

Figure 1(d)

PLIF flow visualization of a shock-accelerated light gas cylinder

J. W. Jacobs

California Institute of Technology

A sequence of fluorescent images is shown, depicting the time evolution of the cross section of a light gas cylinder as it distorts following the passage of a weak shock wave. A round laminar jet is used to produce the cylinder by injecting a 0.79 cm diameter fully developed laminar flow of helium into the initially air-filled GALCIT 17 in. shock tube. Baroclinically generated vorticity, produced by the interaction of a Mach 1.1 shock wave with the density gradient at the jet boundary, causes the circular cylinder to distort initially and then roll up, eventually forming a vortex pair. The laser-induced fluorescence was accomplished by mixing biacetyl, a gaseous fluorescent dye, with the helium stream and illuminating it with a pulsed laser light sheet positioned 2.5 cm from the jet exit. The resulting fluorescent image was captured using an intensified solid-state video camera. The image data is displayed in false color where yellow represents the highest biacetyl concentration and white the lowest. The first photograph in the sequence, Fig. 1(a), is a cross section of the undisturbed cylinder, providing a view of the initial condition. Next are images from three separate runs, Figs. 1(b)–1(d), taken 0.34, 0.49, and 0.79 ms after the passage of the shock wave, which illustrate the transformation of the initially circular cylinder into a pair of vortices.

This work was supported by the Air Force Office of Scientific Research.

Keywords

laser-induced fluorescence; line vortices.

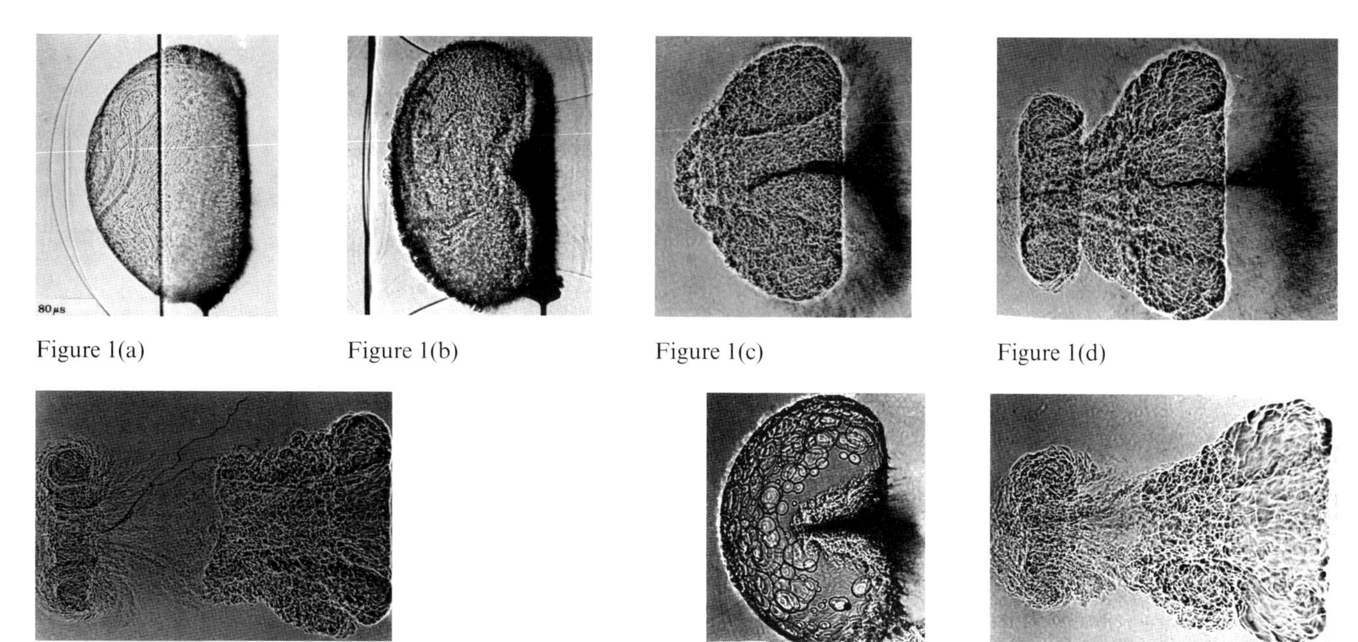

Figure 1(a) Figure 1(b) Figure 1(c) Figure 1(d)

Figure 1(e) Figure 2(a) Figure 2(b)

Shock-induced deformation and mixing of a helium sphere immersed in air

Jean-François Haas and Bradford Sturtevant
California Institute of Technology

The interaction of a plane shock wave with a single spherical gaseous inhomogeneity illustrates the mechanisms by which finite-amplitude waves in nonhomogeneous media generate turbulence and intensify mixing. Shown are two sequences of shadowgraph photographs of shock interaction with a helium-filled soap bubble. The shock propagates from right to left. Figures 1 and 2 show the development of the flow after passage of shocks of strength $M_s = 1.25$ and 1.10, respectively. Figure 1(a) (taken 82µs after the shock hits the right edge of the helium bubble) shows the plane incident and diffracted shock passing outside the bubble and the curved internally refracted wave after it emerges from the bubble. The upstream (right) air–helium interface flattens and forms a reentrant depression [Fig. 1(b), 145µs], which develops into an air jet along the streamwise axis of the helium volume [Figure 1(c), 223µs]. The black cloud to the right of the deforming sphere is an aerosol formed from the soap film. The air jet impinges on the downstream (left) edge of the helium volume, generating a helium vortex ring [Figure 1(d), 600µs] that moves rapidly away from the main tube-like volume of helium [Figure 1(e), 1594µs]. The second series shows the same mixing process occuring at a slower pace for a weaker incident shock: the reentrant air jet [Figure 2(a), 510µs] and the ejected helium vortex ring [Figure 2(b), 2541µs]. The mechanism by which the soap film ruptures appears clearly for this weaker shock: many holes appear simultaneously and grow at the Culick velocity until they merge [Figure 2(a)]. They are much more numerous and therefore remain indistinct for the stronger shock [Figures 1(a) to 1(c)].

The gas flow field is driven by the baroclinic instability: i.e. the deposition of vorticity due to the misalignment of the pressure gradient of the shock(s) and the density gradient of the air–helium curved interface and the subsequent dynamics of this vortex layer,

$$\frac{\partial \vec{\omega}}{\partial t} = \vec{\omega} \cdot \nabla \vec{u} - \vec{\omega} \cdot div \vec{u} + \frac{\nabla \rho \times \nabla p}{\rho^2}.$$

In principle, the velocities of the observed features (upstream and downstream interfaces, tip of the air jet, and centerline of the helium vortex ring) can be obtained from the vorticity field. For a rough estimate of some of these velocities, we used instead the growth rate of the Richtmyer–Meshkov instability (or shock-induced Rayleigh–Taylor instability) for a wavy perturbation of amplitude R and wavenumber $\sqrt{2}R$ (R = radius of the sphere) on a plane interface. The perturbation has the same curvature as the sphere. After a correction for initial compression by the incident shock, linear theory yields the velocity of the "crest" along the axis of symmetry,

$$V_c = V\left[1 + \sqrt{2}\left(1 - \frac{V}{V_s}\right)\frac{\rho_{air} - \rho_{helium}}{\rho_{air} + \rho_{helium}}\right],$$

where V is the velocity of a plane interface after a shock of velocity V_s. For the experiment of Figure 1 (V = 177m/s) the predicted velocity is 253 m/s, while the measured velocity of the upstream interface is 190 m/s during the initial flattening stage and 335 m/s later during the reentrant jet stage. For the experiment of Figure 2 (V = 75 m/s) the predicted velocity is 134 m/s and the measured velocities are 87 and 140 m/s, respectively. In both cases the linear theory is applied well outside its range of validity, so the discrepancy suggests that nonlinear saturation in the flattening stage and cumulation in the jet stage are important in these experiments. The fully developed helium vortex rings at the late stage move at 165 m/s (Fig. 1) and 75 m/s (Figure 2). Their circulation (2.21 and 1.08 m^2/s, respectively) is almost identical to that associated with the motion within the helium "tubes" of the effective pistons represented by the reentrant air jets.

Keywords

helium-filled soap bubble; aerosol; Rayleigh–Taylor instability; Richtmyer–Meshkov instability shadowgraph.

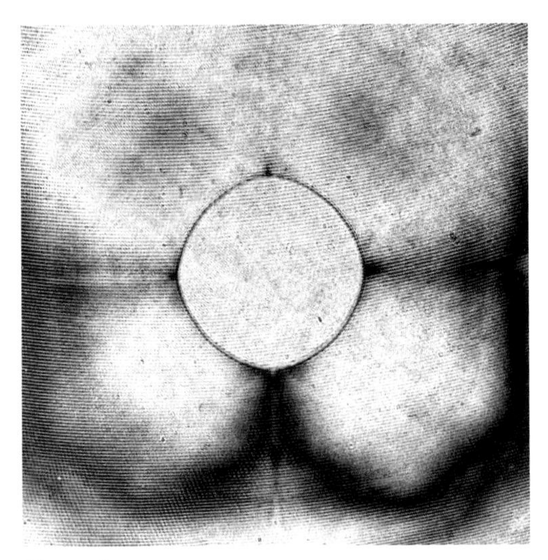

Figure 1

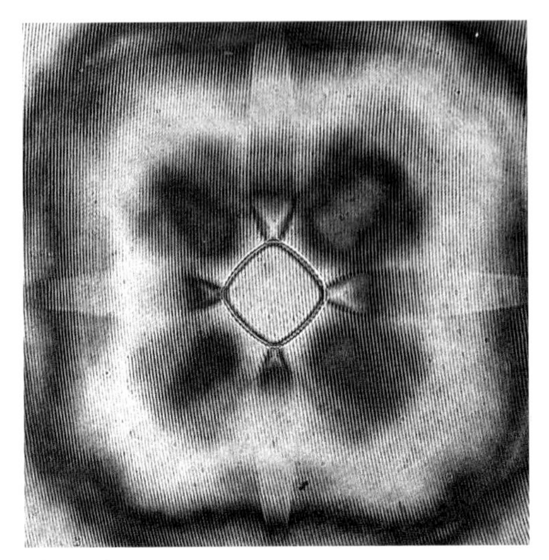

Figure 2

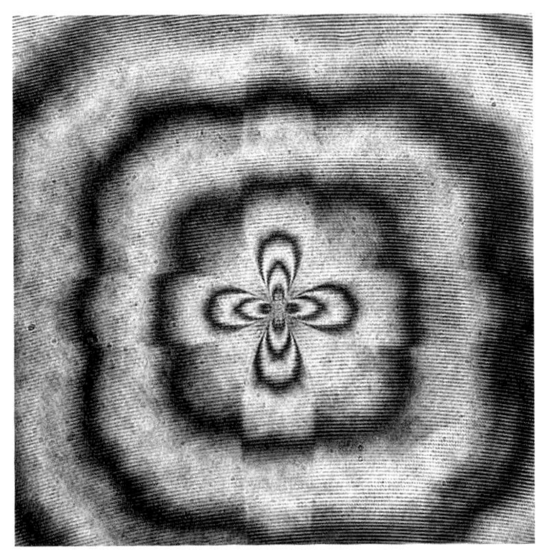

Figure 3

Figure 4

Converging cylindrical shock wave

M. Watanabe and K. Takayama

Tohoku University

Converging toward the center, a cylindrical shock wave becomes unstable. Experimentally, to examine this mechanism, holographic interferometric observation has been conducted. A converging cylindrical shock wave was made by turning a ring-shaped shock wave by 90° in an annular shock tube. Four sequential interferograms show how an initially cylindrically shaped shock deforms when converging toward the center. The initial shock Mach number was 2.0 in air. Flow visualization was carried out by means of double-pulse holographic interferometry.

Although the initial shape was perfectly cylindrical, it was slightly deformed into a quasi-square (Fig. 1). This was attributable to the effect of the four struts that were supporting the inner core of the annular shock tube. The effect of the struts seemed to be too small to disturb initial cylindrical shock configuration, however, these linear perturbations grew and from a certain stage the continuous density profile behind the converging quasicylindrical shock wave failed to exist and resulted in the discontinuous distribution (Fig. 2). The four pairs of the Mach reflection appear (see Fig. 3). The fringes are corresponding to isopicnics and their pattern reminds one of a four-leaved clover. The triple points of these Mach reflections accompany the vortices and reflected shock waves. Very close to the center where the triple points meet, these Mach reflections were the so-called inverted Mach reflection. After collapsing at the center, the shock wave, which was dragging complicated wave interaction behind it, was reflected and became a stable diverging shock wave. The cross-shaped pattern near the center was the remaining vortices and they looked like mushrooms (Fig. 4). Each image covers approximately 58×58 mm^2 area.

Keywords

hologram; shock tube; Mach reflection; shock triple point.

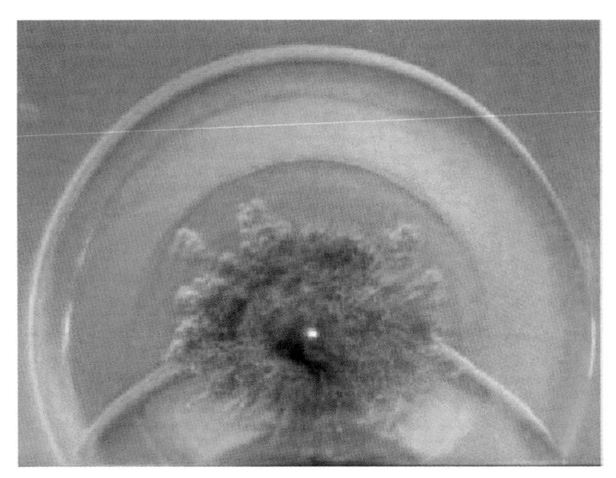

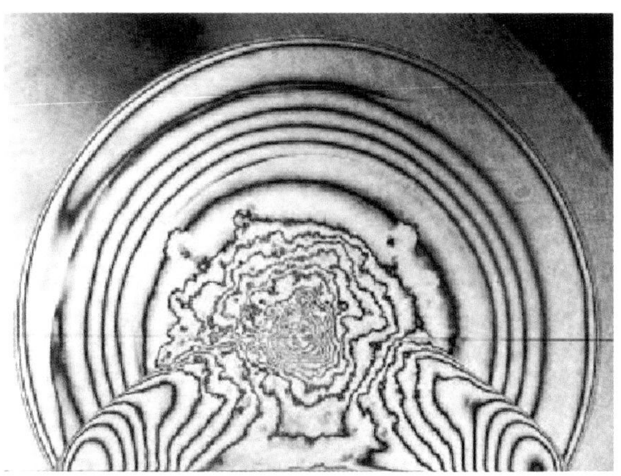

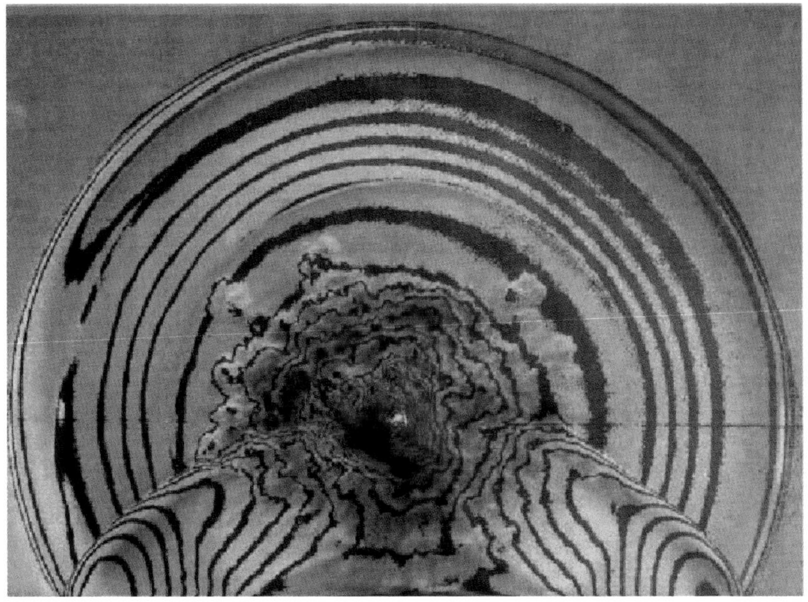

Figure 1

Combined schlieren and interferometry visualization of blast waves

Harald Kleine and Kazuyoshi Takayama

Tohoku University

Density-sensitive methods to visualize compressible fluid flows can be classified into three different groups, depending on the derivative of the density that they record (shadow, schlieren, and interferometry). Each method has its inherent strengths and shortcomings, where the latter may sometimes cause a misinterpretation of the observed flow phenomena. For a complete and unambiguous investigation of a compressible flow it is therefore desirable to apply more than just one visualization technique. In flows with a low degree of reproducibility, this multiple visualization has to be conducted simultaneously, i.e., in the same experiment and at the same instant.

One possible combination of visualization methods consists of a holographic interferometer and a colour schlieren apparatus. This system yields two records of the flow, one interferogram, in which the density distribution is represented by modulations of intensity (fringes), and a schlieren record, which displays the density gradient (here: the direction of the gradient) by different hues. The two images can be analyzed individually or superimposed.

An example of this combined visualization is shown above: a small explosive charge (10 mg of silver azide, AgN_3), suspended by a thin nylon thread 30 mm above a rigid wall and ignited by the pulse of an Nd:YAG laser, generates a blast wave that reflects from the wall. The visualization shows the primary incident and reflected waves as well as the cloud of combustion products and the secondary wave, which at the shown instant has also reflected from the wall. The image is taken 120 μs after ignition close to the point where the reflection pattern of the primary wave changes from regular reflection to Mach reflection.

Keywords

density distribution; density gradient; Mach reflection.

Keyword Index

acoustic forcing 5, 27, 85
acoustic wave 110
aerosol 113
air bubbles 52
air entrainment 75, 77
airfoil 41
atomization 39, 73

barrel shock 111
bluff body 89
bluff-body wake 21, 22
Boger fluid 62
boundary-layer detachment 107
boundary-layer ejection 10
boundary-layer instability 36, 101
boundary-layer transition 99
bubble collision 52
bubble eruption 58
bubble formation 77
buoyancy convection 33

capillary length-scale 51
capillary pinchoff 47
capillary wave 51, 77, 102
cavitation bubble 43
cavity collapse 75
centrifugal instability 26
Cessna aircraft 19
chaotic mixing 31, 45
chaotic oscillations 32
chemical reaction 27
chimneys 34
clustering 57
comb teeth 103
convection cells 100
convective instability 94
Couette device 59
counter-rotating vortices 90

dendrites 34
density distribution 115
density gradient 115
direct image-transfer 29
downwash far-field 19
dripping faucet 44
droplet breakup 45
droplet ejection 39, 47
droplet formation 49
droplet pinchoff 44
dust resuspension 61
dye visualization 4, 15, 26

electric field 100
electrostatic spray 49
evaporation 94
extensional flow 62

far wake 98
Faraday forcing 39, 47, 55, 76
film jet 106
finger convection 34
finger selection 95
fingers 70
flake reflection 36, 96
flame extinction 84
flame impingement 86
fluid filament 62
fluorescein 65
fluorescent dye 16, 30, 38, 74, 80, 90, 102
forced jet 73
fractal dimension 68
free shear layer 2, 5

granular mixing 56
gravitational fingering 68
gravity 75
gravity wave 102

hairpin vortex 74
heat-transfer patterns 105
Hele-Shaw cell 68, 71, 95
helium-filled soap bubble 113
hologram 41, 114
horseshoe vortex 7
hydrogen–helium flame 85

immiscible displacement 70
inertia 69

inertial oscillations 51
interfacial tension 45
interferometry 32, 41

jet atomization 46
jet breakup 79
jet collisions 65
jet impingement 73

Kelvin–Helmholtz instability 48, 89
Klebanoff breakdown 99

laminar mixing layer 27
laminar–turbulent transition 83, 101
laser sheet 16, 22
laser-induced fluorescence 2, 14, 18, 22, 23, 91, 112
Leidenfrost effect 43
ligament breakup 64
line vortices 112
liquid chain 64, 65
liquid dye 24, 31
liquid sheet disintegration 48
liquid-metal jet 79

Mach disk 111
Mach reflection 114, 115
maelstrom 40
Marangoni stresses 38, 67
methane flame 86, 87
microgravity 71
Mie scattering 8, 27, 83, 84, 85, 87, 89
mixing transition 2
mushroom mode 93
mushroom vortex 4, 74, 78

nematic liquid crystal 100
noise generation 109

oil paint visualization 29
oscillatory shear 37

particle jetting 58
particle mixing 56
particle ripples 35
particle splash 58
passive tracer 30
pathlines 17
pattern selection 33
periodic vortex rings 90
plume 94, 104
Poincaré map 12
premixed flame 82

Rayleigh instability 76
Rayleigh scattering 111
Rayleigh–Bénard convection 104, 107
Rayleigh–Taylor instability 50, 91, 93, 113
reflected shock 111
resonance forcing 9
Richtmyer–Meshkov instability 91, 93, 113
rotating cylinders 96
round jet 4, 109

Saffman–Taylor instability 95
salt cubes 56
sand ripples 37
schlieren 82
shadowgraph 32, 33, 113
shear layer 4, 36, 106
shear-induced segregation 59
shedding waves 98
shock noise 110
shock structure 79
shock triple point 111, 114
shock tube 93, 114
simulation 9, 12, 13, 17, 25, 35, 60, 92, 109, 110
simulation and experiment 44
sinuous mode 93
smoke visualization 5, 7, 10, 20, 98, 99
soap film 3, 21, 103
spiral mode instability 85
spray generation 46, 49, 64
spreading 67
start-up vortex 13, 16
steel balls 57
streaklines 92
streamline 87
streamtube 87
stretched flame 84
stretched vortex 26
stretching and folding 30
subharmonic forcing 55
surface pressure 17
surface tension 43, 44, 66, 69, 75
surface-tension gradients 38
swirl 104

Taylor–Couette cell 96
thermal plumes 107
thermochromic liquid crystals 104, 105, 107
tidal vortex 40
time-periodic forcing 30
Tintenpilze 60
Tollmien–Schlichting wave 99
toroidal vortices 83
tow tank 23
triple exposure 53
turbine cascade 105
turbulent wake 18
two-color PIV 106

unstable manifold 12
UV-light 16

vapor bubble 50
vibrating plate 55, 57
viscous fingering 67, 71
vortex braids 6, 18, 80
vortex breakdown 14
vortex bubble 14, 20
vortex collision 24
vortex dipole 24
vortex dynamics 5
vortex ejection 9
vortex generation 15, 103
vortex instability 15

vortex loop 22
vortex pairing 2, 8, 10, 25
vortex pairs 18
vortex ring 6, 7, 53, 60, 80
vortex shedding 13, 23
wake vortex 7, 61
water bells 69
water wave 102
wave breaking 102
whirlpool 40